Auditing Artificial Intelligence

Artificial Intelligence (AI) is revolutionizing industries, yet its rapid evolution presents unprecedented challenges in governance, ethics, and security. *Auditing Artificial Intelligence* is an essential guide for IT auditors, information security experts, and risk management professionals seeking to understand, evaluate, and mitigate AI-related risks.

This book provides a structured framework for auditing AI systems, covering critical areas such as governance, compliance, algorithm transparency, ethical accountability, and system performance. With 24 insightful chapters, it explores topics including:

- **AI Governance and Ethics** – Establishing frameworks to ensure fairness, accountability, and transparency in AI deployments.
- **Risk Management and Compliance** – Addressing the legal and regulatory landscape, including GDPR, the EU AI Act, and ISO standards.
- **Bias and Trustworthiness** – Evaluating AI decision-making to detect bias and ensure equitable outcomes.
- **Security and Continuous Monitoring** – Safeguarding AI systems from adversarial attacks and ensuring operational consistency.
- **Model Performance and Explainability** – Assessing AI outputs, refining accuracy, and ensuring alignment with business objectives.

Designed for professionals tasked with assessing AI systems, this book combines practical methodologies, industry standards, and real-world audit questions to help organizations build responsible and resilient AI practices and assess associated risks. Whether you are assessing AI governance, monitoring AI-driven risks, or ensuring compliance with emerging regulations, this handbook provides the guidance you need to navigate and assess the complexities of AI systems with confidence.

Stay ahead in your role and responsibility for assessing the rapidly evolving deployment and use of AI across the organization – equip yourself with the knowledge and tools to ensure its responsible, safe, approved, secure, and ethical use.

Auditing Artificial Intelligence

A Handbook for Audit, Risk, and Security Professionals

Albert J. Marcella

CRC Press
Taylor & Francis Group
Boca Raton London New York

CRC Press is an imprint of the
Taylor & Francis Group, an **informa** business

Designed cover image: Shutterstock Image ID 2347708781

First edition published 2026
by CRC Press
2385 NW Executive Center Drive, Suite 320, Boca Raton FL 33431

and by CRC Press
4 Park Square, Milton Park, Abingdon, Oxon, OX14 4RN

CRC Press is an imprint of Taylor and Francis Group, LLC

ISBN: 978-1-041-07879-1 (hbk)
ISBN: 978-1-041-08675-8 (pbk)
ISBN: 978-1-003-64650-1 (ebk)

DOI: 10.1201/9781003646501

Typeset in Sabon
by codeMantra

To Diane….

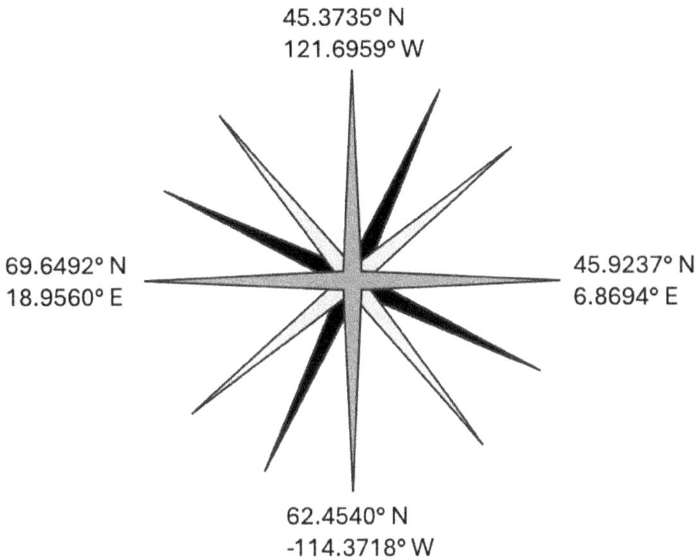

My wife, best friend, and partner in every adventure.

You fill my days with happiness, my life with love.

Let us wander together, hand-in-hand, forever among the stars.

$\infty + 1$

Contents

Figures and Tables

Figures

Tables

Foreword

I am a strong believer that advanced technologies like Artificial Intelligence (AI) have the potential to transform our lives and society dramatically and positively. However, like all technologies throughout history, AI is a double-edged sword, offering tremendous benefits and posing significant risks if not managed and used responsibly.

"Technology is a useful servant but a dangerous master," cautioned Norwegian historian Christian Lous Lange, a warning that resonates deeply in the age of AI. As AI transitions from a futuristic vision to a transformative force – redefining industries, economies, and societies – the urgency for robust oversight has never been greater. With their remarkable ability to process vast data, make autonomous decisions, and fuel innovation, these systems weave a complex tapestry of opportunities and risks. Yet, as organizations rush to harness AI's power, a pressing question emerges: How do we ensure these tools remain our servants and not our masters? The answer lies with those entrusted to uphold integrity – audit, risk, and security professionals. For these guardians of accountability, *Auditing Artificial Intelligence: A Handbook for Audit, Risk, and Security Professionals* stands as an indispensable guide.

This handbook arrives at a critical juncture in the AI age. AI's swift evolution – from generative models producing human-like content to reinforcement learning systems optimizing complex operations – has outpaced the development of standardized tools for its scrutiny. Traditional audit methodologies, while essential, often falter when applied to the dynamic, opaque, and data-driven nature of AI. Stepping into this gap, Dr. Al Marcella, with nearly five decades of expertise in IT audit, risk management, and cybersecurity, delivers a groundbreaking resource that marries theory with practice in a way that is both accessible and profoundly useful.

What sets this book apart is its practical, hands-on approach. Within these pages, readers will discover over 1,300 meticulously crafted, open-ended audit questions spanning more than 30 operational areas of AI. These are not mere checklists; they are instruments of inquiry, designed to peel back the layers of AI systems and expose their strengths, vulnerabilities, and alignment with organizational and ethical imperatives. Whether you are evaluating the fairness of a machine learning algorithm, the robustness of a generative AI tool, or the compliance of an AI deployment with global standards, this handbook provides a roadmap to navigate these challenges with confidence.

Dr. Marcella's credentials lend undeniable authority to this work. As the founder of Business Automation Consultants, LLC, he has spent decades at the forefront of IT governance. His accolades – such as the Information Systems Security Association's Security Professional of the Year (2016) and the Institute of Internal Auditors' Leon R. Radde Educator of the Year (2000) – reflect a career defined by innovation and impact. With 30 books, numerous articles, and a global presence as a speaker, Dr. Marcella has consistently anticipated the profession's needs and delivered enduring solutions. From auditing multinational corporations to exploring cyber forensics and emerging technologies, his journey infuses this handbook with unparalleled expertise.

This is more than a technical guide; it is a call to action. It empowers professionals to move beyond compliance checklists and embrace a deeper, more inquisitive approach to AI oversight. In doing so, it addresses not only the "what" of AI auditing – its processes and metrics – but also the "why": the ethical, societal, and operational stakes at play. As AI permeates sectors from healthcare to finance to public policy, the demand for such a resource will only grow. Auditors, risk managers, and security experts will find in these pages a trusted companion to meet that demand head-on.

I invite you to explore this book with an open mind and a critical eye. Let it challenge your assumptions, sharpen your skills, and inspire you to ask the questions that matter most. In a world increasingly shaped by artificial intelligence, the responsibility of ensuring its integrity rests with us all, and with this book, we are well-equipped to rise to the occasion.

Alan Tang, Ph.D., CISSP, CISA
April 14, 2025

Preface

After 46 years of working in the information technology (IT) profession, from programming mainframe computers to developing and delivering IT audit and security training programs, I have relied on what the profession affectionately calls an "audit toolbelt" when conducting IT audits and risk assessments.

An "audit toolbelt" refers to the collection of skills, techniques, technologies, and resources that auditors use to conduct effective audits. Just like a physical toolbelt contains various tools for different tasks, an audit toolbelt encompasses all the essential elements to assess, analyze, and evaluate organizational processes, controls, and risks.

One tool I often found missing from this "toolbelt," and more importantly, since I was tasked with auditing the ever-evolving technology underlying IT environments, were predefined templates for IT audit plans, risk assessments, and associated checklists to ensure critical areas are covered during the audit process. Due to the sometimes-unique nature of an IT audit engagement, developing these checklists, often from scratch, took valuable time from the engagement.

One of the many challenges facing professionals responsible for auditing IT environments and assessing risks associated with the broad spectrum of technologies inherent in most enterprise operations has always been creating targeted, open-ended questions to assess the existence (or lack thereof) of functioning internal controls. Internal controls that have been specifically developed and implemented to ensure the integrity, efficiency, and effectiveness of an organization's IT operations and platforms while safeguarding its assets and resources. These controls are essential for achieving organizational objectives and minimizing risks.

In the context of performing an audit, open-ended questions are inquiries that require respondents to provide detailed, elaborate, explanatory, or narrative answers rather than just simple "yes" or "no" one-word responses. These questions typically begin with words like "how," "what," "why," "describe," or "tell me about," encouraging the interviewee to share comprehensive information, explanations, and insights.

Open-ended questions allow auditors (and risk assessment professionals) to delve deeper into processes, controls, and procedures. They enable these professionals to understand not just what is done but how and why it is done. Open-ended responses can provide information necessary to identify the root causes of issues, not just the symptoms. They can also help in uncovering problems or risks that might not surface through direct questioning.

When performing an IT audit, be it of a financial application, a network, or a business resiliency plan, for example, I had often thought that having a handbook, and a database, of open-ended questions, which I could query, select, and customize for the current audit engagement would have greatly assisted me in the development of an initial audit questionnaire, and a comprehensive audit program, saving both time and money in the long run.

The need for a "go-to" resource, a handbook of open-ended questions on auditing artificial intelligence (AI) and finding none currently available led me to develop this book...*Auditing Artificial Intelligence: A Handbook for Auditors and Risk Professionals*©.

A "handbook" is typically a concise, practical guide designed to provide essential information, instructions, or guidelines on a particular subject. Unlike a full-length book, a handbook is usually more focused and straightforward, often written in a way that allows users to easily find the specific information they need without extensive reading.

Handbooks can serve various purposes, such as:

1. Instructional Manual – Offering step-by-step guidance or processes.
2. Reference Guide – Serving as a quick reference for specific terms, concepts, or practices.
3. Smart Practices – Outlining vetted practices, tips, or standards in a given field.

This book is based on these concepts and has been written to be a user-friendly, go-to resource for practical knowledge or skills. A resource for those individuals tasked with auditing and assessing internal controls, risks, and security of AI systems and environments. Additionally, the book serves as an excellent resource for individuals desiring to gain a better, overall understanding and insight into the rapidly evolving field of AI.

Keeping with this definition, the reader will find in this handbook over 1,300 open-ended audit questions covering over 20 AI operational areas. These questions were specifically written as open-ended questions, lending them to easily be used in developing an initial audit checklist and can be used "as is" or customized to fit the reader's specific AI audit or risk assessment requirements.

I trust that this handbook will assist audit, risk and security management professionals in the timely construction of AI audit questionnaires critical and essential for the continuous audit and assessment of AI and, frankly, any professional who is tasked with managing the many diverse facets of AI development and deployment.

Acknowledgements

The author would like to thank the many contributors who have provided input to this book and the multitude of researchers, authors, AI system developers, global government agencies, professional associations, and data scientists whose research, white papers, articles, presentations, and books provided a rich, informative, and educational basis for both the approach and eventual development of this handbook.

The author also wishes to thank the anonymous reviewers identified by the Publisher, CRC Press, who initially reviewed this book and provided positive feedback and comments on the proposed content when the author initially brought the idea to CRC Press.

The author would like to personally thank the following professionals for their insights and comments on the proposal concept for this book.

Mr. John G. Tannahill, CISM, CRISC
President/CEO
J. Tannahill and Associates

Mr. John Kitchen
Cybersecurity Leader

Dr. Ali Ovlia
Professor
Department of Computer Science
The George Herbert Walker School of Business and Technology
Webster University

Mr. Douglas Menendez, CISA
Adjunct Professor
Department of Computer Science
Webster University

Examples Are Not Endorsements

This book contains references to materials that are provided for the reader's convenience. The inclusion of these references is not intended to reflect their importance, nor is it intended to endorse any views expressed or products or services offered by third-party providers. These reference materials may contain the views and recommendations of various subject matter experts as well as hypertext links, contact addresses, and websites to information created and maintained by other public and private organizations.

The opinions expressed in any of these third-party materials included as references or examples do not necessarily reflect the positions or policies of the author.

The author does not control or guarantee the accuracy, relevance, timeliness, update, or completeness of any outside, third-party information included in this book. These references should not be construed as an endorsement by the author or by the publisher. The reader should validate and substantiate any information directly from third-party providers before authorizing the acquisition, implementation, or use of any product or service referenced in this book.

Author Biography

Albert (Al) J. Marcella, Ph.D., CISA, CISM, President of Business Automation Consultants (BAC) LLC, is an internationally recognized public speaker, researcher, IT consultant, and workshop and seminar leader with 46 years of experience in IT audit, risk management, IT security, and assessing internal controls, having authored numerous articles and 30 books on various IT, audit, and security related subjects. Dr. Marcella's clients include organizations in financial services, IT, banking, petrolchemical, transportation, services industry, public utilities, telecommunications, and departments of government and nonprofits.

Research conducted by Dr. Marcella on unmanned aircraft systems, cyber extortion, workplace violence, personal privacy, electronic stored information, privacy risk, cyber forensics, disaster and incident management planning, the Internet of Things, ethics, and astrophotography has been published in the *ISACA Journal*, *Disaster Recovery Journal*, *Journal of Forensic and Investigative Accounting*, *EDPACS*, *ISSA Journal*, *Continuity Insights*, *Internal Auditor Magazine*, and the *Astronomical League's Reflector Magazine*.

Dr. Marcella, along with co-authors Madeline Parisi and Brian Moore's four-book series *From Street-smart to Web-wise®: A Cyber Safety Training Program Built for Teachers and Designed for Children,* released in 2025 by Taylor and Francis Publishing provides K-8 educators, parents, and educators globally, who nurture, guide, support, and ignite a passion for knowledge and learning in children of every educational background, with guidance and tools needed to make student cyber-safety awareness practical, fun, and impactful.

Dr. Marcella holds a B.S. degree in Management, a B.S. degree in Information Technology Management, an MBA with a concentration in Finance, and a Ph.D. in Management/Information Technology Management. Dr. Marcella is a Certified Information Systems Auditor (CISA), a Certified Information Security Manager (CISM), and holds an ISACA Cybersecurity Certificate.

Dr. Marcella is the 2016 recipient of the Information Systems Security Association's Security Professional of the Year award and recipient of the Institute of Internal Auditors Leon R. Radde Educator of the Year 2000 award and has been recognized by the Institute of Internal Auditors as a Distinguished Adjunct Faculty Member. Dr. Marcella has taught IT audit seminar courses for the Institute of Internal Auditors (IIA) and the Information Systems Audit and Control Association (ISACA).

Introduction

The rapid growth, development, deployment, acceptance, implementation, and use of artificial intelligence (AI) across the enterprise has created an urgency and a realization that an audit (performed internally or via an external third party) of AI systems development, platforms, applications, and the risks associated with AI can no longer be ignored or delayed.

Continuously auditing AI systems is critical for organizations due to the dynamic and evolving nature of these systems and their impact. An organization's stakeholders, along with society's acceptance of an organization's safe and ethical use of AI will depend greatly on the performance of independently directed AI audits. Audits that, in some cases, are legally (contractually) or legislatively mandated and, in all cases, will require the organization's immediate attention and Board and executive-level management commitment and support.

By auditing and validating the existence of a robust system of established and functioning internal controls, specifically addressing AI, organizations can create a secure, efficient, and trustworthy environment in which AI plays an essential role in supporting the organization's mission and objectives.

The material in the following chapters has been designed and written to assist auditors, risk management, and security professionals in developing a useful and practical AI audit or risk assessment program. Each chapter provides numerous open-ended questions related to a specific AI operational/functional topic.

The reader is encouraged to review the entirety of the questions presented for each topic, then select specific questions to incorporate into an audit checklist/program, and, where necessary, customize the questions for the reader's operating environment and audit/assessment requirements.

As with any book written on technology, and especially a technology as dynamic, forward-moving, and transformative as AI, the AI areas identified in this book and recommended for audit examination and risk assessment will inevitability change over time as AI technology advances and as an organization's use of AI changes to keep pace with these advancements.

The open-ended audit questions presented throughout this *Handbook* are not exhaustive; the questions and the issues they are designed to assess are intended as a starting point for further exploration and development.

That fact established the specific AI areas that follow, and the associated audit/risk assessment questions provided for each area are done so with the full knowledge that they are representative of what certainly will be an evolving technology... a technology that will require a continuous audit and risk assessment presence.

The Audit of Artificial Intelligence

Audit Objective

To ensure that AI systems employed by the organization are effectively managed, secure, compliant, and aligned with business objectives by evaluating governance and strategy, assessing implementation and integration, ensuring data management and privacy, reviewing security and risk management, evaluating internal controls, verifying compliance, monitoring performance, promoting ethical and responsible AI use, engaging stakeholders, and providing recommendations for improvement. This comprehensive audit aims to maintain trust, security, and compliance while supporting the organization's strategic goals and continuous improvement in AI system management and use.

Audit Scope

Evaluate and assess whether the design, development, acquisition, deployment, and operating environment of AI tools, which are currently operating within the organization, are performing to expectations and operating within the stated policy and procedure limitations and guidelines as established by appropriate regulatory compliance requirements and as approved by management.

This includes assessing the AI algorithms for accuracy, transparency, accountability, non-discrimination, and compliance with relevant regulations and standards, and that algorithms perform correctly, fairly, and ethically within their intended operational environments.

The audit will also seek to identify any risks or issues related to data handling, model training, output interpretation, and decision-making processes and to verify that appropriate governance, oversight, and control mechanisms are in place.

Audit Approach

To accomplish the stated objectives of an IT audit of AI, the approach to the performance of the audit should include the following steps:

1. **Engage Stakeholders:** Involve key stakeholders, including management, IT staff, data scientists, and end-users, to gather insights,

identify areas of concern, and ensure alignment with business objectives.
2. **Conduct Risk Assessment:** Perform a comprehensive risk assessment to identify potential threats, vulnerabilities, and compliance requirements, focusing on data management, security, and regulatory compliance.
3. **Evaluate Governance and Strategy:** Assess the AI strategy and governance framework to ensure they align with organizational goals and include effective policies and procedures for AI development, deployment, and use.
4. **Assess Implementation and Integration:** Verify the effective implementation and integration of AI systems within the existing IT infrastructure, ensuring accurate and reliable data inputs and outputs.
5. **Review Internal Controls:** Evaluate the adequacy and effectiveness of internal controls related to AI systems, including access controls, monitoring, and logging to prevent and detect errors, fraud, and misuse.
6. **Ensure Data Management and Privacy:** Review data governance practices to ensure data quality, integrity, and security, and compliance with data privacy regulations.
7. **Monitor Performance:** Assess performance-monitoring mechanisms to ensure AI systems meet expected criteria and evaluate ongoing maintenance and update procedures.
8. **Promote Ethical AI Use:** Ensure AI systems are used ethically, with considerations for fairness, transparency, and accountability, preventing unintended biases or discriminatory outcomes.
9. **Provide Recommendations:** Document findings and provide actionable recommendations for improvements to enhance AI system management, security, compliance, and alignment with strategic goals. Regular reviews and updates to the audit scope will ensure it remains relevant and addresses emerging issues effectively.

Figure 0.1 provides the reader with an example of an AI Audit engagement timeline. As noted, the task duration and completion metric (days, weeks, months) is highly dependent not only upon the scope of the audit as decided by management but, additionally on the number of audit professionals assigned to the engagement and the skillset of those assigned.

Audit Team

The audit team ideally will consist of professionals with backgrounds that include data science, information technology, networks, legal, compliance, cybersecurity, internal controls, ethics, and domain-specific knowledge to conduct the audit both comprehensively and effectively.

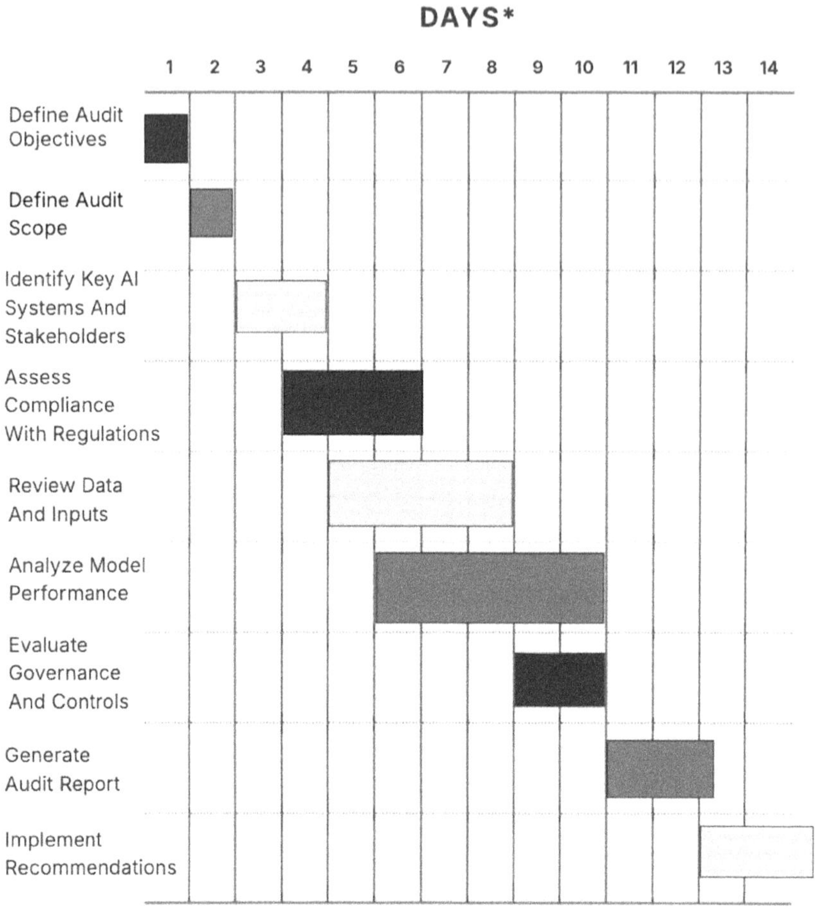

* Timeline (days, weeks) is organization and engagement specific. Task duration and completion will depend heavily on the number of auditors assigned to the engagement and each auditor's skill set relevant to AI and IT in general.

Figure 0.1 Auditing AI – Engagement Timeline (Example)[1]

Auditor Skill Set Requirements

Auditors need specific training in the following areas of expertise to perform meaningful assessments of AI applications and to give appropriate recommendations:

- Auditors need a good understanding of the high-level principles of ML algorithms and up-to-date knowledge of the rapid technical developments in this field – this is sufficient to perform a baseline audit by reviewing the respective documentation of an ML system.

- For a thorough audit that includes substantial tests, auditors need to understand common coding languages and model implementations and be able to use appropriate software tools.
- ML-related IT infrastructure often includes cloud-based solutions due to the high demand for computing power. Therefore, auditors need a basic understanding of cloud services for this type of audit work.[2]

See Appendix 1 for additional information on audit objectives, scope, and general AI audit questions.

REFERENCES

(n.a.), (n.d.), "Machine Learning Glossary," Google, https://developers.google.com/machine-learning/glossary.

(n.a.) 2019, "OECD CAI Recommendation," Recommendation of the Council on Artificial Intelligence, OECD, OECD Legal Instruments, https://legalinstruments.oecd.org/en/instruments/oecd-legal-0449.

(n.a.), (n.d.), "Neural Networks," National Library of Medicine, www.nnlm.gov/guides/data-glossary/neural-networks.

(n.a.), (n.d.), AI Fairness 360, https://aif360.mybluemix.net/resources#glossary.

(n.a.), (n.d.), APA Dictionary of Psychology, American Psychological Association (APA), https://dictionary.apa.org/ethics.

(n.a.), (n.d.), IAPP Privacy Glossary, Glossary of Privacy Terms, https://iapp.org/resources/glossary.

(n.a.), (n.d.), "What are recurrent neural networks?" IBM, www.ibm.com/topics/recurrent-neural-networks.

(n.a.), 2015, "Technology Investment Strategy 2015-2018," United States Department of Defense's Test and Evaluation, Verification and Validation (TEVV) Working Group, https://defenseinnovationmarketplace.dtic.mil/wp-content/uploads/2018/02/OSD_ATEVV_STRAT_DIST_A_SIGNED.pdf.

(n.a.), 2017 "IEEE Guide for Terms and Concepts in Intelligent Process Automation," in IEEE Std 2755-2017, vol., no., pp.1-16, 28, doi: 10.1109/IEEESTD.2017.8070671.

(n.a.), 2021, "National Security Commission on Artificial Intelligence: The Final Report," www.nscai.gov/wp-content/uploads/2021/03/Full-Report-Digital-1.pdf.

(n.a.), 2022, Operations Research and Analytics, INFORMS, www.informs.org/Explore/Operations-Research-Analytics.

(n.a.), December 23, 2022, "Chapter 1: Subject Matter and Scope of Copyright, Section 101 Definitions" Copyright Law of the United States (Title 17), U.S. Copyright Office, www.copyright.gov/title17/92chap1.html.

(n.a.), January 2023, "Artificial Intelligence Risk Management Framework (AIRMF1.0)," NIST, https://nvlpubs.nist.gov/nistpubs/ai/NIST.AI.100-1.pdf.

Chandler, D., Munday, R., 2011, A Dictionary of Media and Communication (1 ed.), Oxford University Press, ISBN-13: 9780199568758.

ChaptGPT, May 31,2023, "What is Transformer Architecture technology?" https://chat.openai.com/share/44f518e0-944f-4765-8b19-36377d6d63c5.

GWUC Cyber Glossary, 2018, The George Washington University, Natural Security Archive - The Cyber Vault Project, https://nsarchive.gwu.edu/cyber-glossary-b.
High-Level Expert Group on AI (AI HLEG), April 8, 2019, "Ethics Guidelines for Trustworthy AI," https://ec.europa.eu/futurium/en/ai-alliance-consultation.1.html.
High-Level Expert Group on AI (AI HLEG), April 8, 2019, A Definition of AI: Main Capabilities and Disciplines, https://ec.europa.eu/futurium/en/ai-alliance-consultation.1.html.
Hutson, M., 2017, "AI Glossary: Artificial intelligence, in so many words," Science, Vol 357, Issue 6346, p. 19, www.science.org/doi/10.1126/science.357.6346.19.
Information Technology Laboratory Computer Security Resource Center (CSRC) Glossary, https://csrc.nist.gov/glossary.
Lutkevich, B. June 27, 203, "Artificial intelligence glossary: 60+ terms to know," TechTarget, https://tinyurl.com/49wap97b.
NIST Special Publication 800-37 Revision 2, Risk Management Framework for Information Systems and Organizations: A System Life Cycle Approach for Security and Privacy, https://csrc.nist.gov/pubs/sp/800/37/r2/final.
National Security Commission on Artificial Intelligence (NSCAI), Appendix A: Technical Glossary, www.nscai.gov/wp-content/uploads/2021/03/Full-Report-Digital-1.pdf.
O'Brien, S., June 7, 2022, "7 Key Differences Between NLP and Machine Learning (and Why You Should Learn Both)," Marketing AI Institute, www.marketingaiinstitute.com/blog/7-key-differences-between-nlp-and-machine-learning-and-why-you-should-learn-both#:~:text=Machine%20learning%20is%20a%20subset,learning%20to%20be%20used%20effectively.
Pasick, A., March 27, 2023, "Artificial Intelligence Glossary: Neural Networks and Other Terms Explained," www.nytimes.com/article/ai-artificial-intelligence-glossary.html.
Sánchez, R., González, F., Ortiz, T., December 2020, "Responsible use of AI for public policy: Data science toolkit," Inter-American Development Bank (IDB), Organisation for Economic Co-operation and Development (OECD), https://publications.iadb.org/en/responsible-use-ai-public-policy-data-science-toolkit.
Tabassi, E., Burns, K., Hadjimichael, M., Molina-Markham, A., Sexton, J., October 2019, "A Taxonomy and Terminology of Adversarial Machine Learning," Draft NISTIR 8269, https://doi.org/10.6028/NIST.IR.8269-draft.
The Organisation for Economic Co-operation and Development (OECD), February 2023, "Advancing accountability in AI: Governing and managing risks throughout the lifecycle for trustworthy AI", OECD Digital Economy Papers, No. 349, OECD Publishing, Paris, https://doi.org/10.1787/2448f04b-en.

NOTES

1 Marcella, A., December 18, 2024, Figure 0.1, Auditing AI – Engagement Timeline (Example), drawn by the author.
2 (n.a.), April 27, 2023, "Auditing machine learning algorithms," The Supreme Audit Institutions of Finland, Germany, the Netherlands, Norway and the UK, www.auditingalgorithms.net/

Part I

Auditing Artificial Intelligence

Chapter 1

Artificial Intelligence

The Basics

> "The real risk with AI isn't malice, but competence. A super-intelligent AI will be extremely good at accomplishing its goals, and if those goals aren't aligned with ours, we're in trouble."
>
> – Stephen Hawking

INTRODUCTION

Artificial intelligence (AI) in general and the now popular "generative AI" sometimes called generative AI, in particular, such as Open AI's ChatGPT[1] and DALL-E,[2] and others, including Microsoft CoPilot and Google Gemini, are receiving increasing coverage in the media and escalating interest from the general public. While interest in AI continues to grow, the reality today is that many professionals do not fully understand or have a basic insight into or familiarity with the fundamental concepts and working principles of AI or generative AI.

The material presented in this chapter is designed to provide the reader, especially a reader new to the field of AI, an introduction, albeit a brief and broad one, to the basics of AI. Beginning with defining AI to examining the component pieces that make AI...well, AI.

AI, if not part of the organization's business model now, will be within the next three to five years, if not sooner (see Figure 1.1).

DEFINING ARTIFICIAL INTELLIGENCE

There is no universally agreed-upon definition of AI due to its wide variety of ways to support, enhance, and automate human activities and its ability to learn and act autonomously.

Defining AI would require one to first ask, "What is intelligence?"

DOI: 10.1201/9781003646501-2

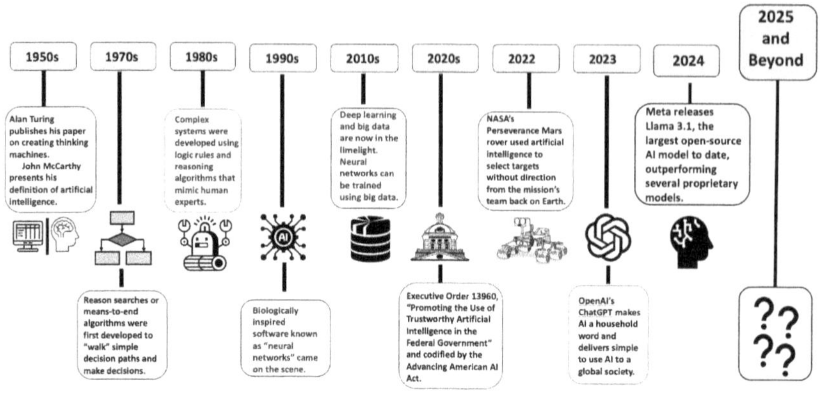

Figure 1.1 Timeline of Select Artificial Intelligence Milestones.[3]

Then, one has to examine and subsequently define the various elements that comprise the environment in which AI functions and operates (data science), such as computational logic, AI algorithms, machine learning (ML), deep learning (DL), neural networks, large language models, probabilistic reasoning, training data, optimization techniques, inference, and, of course, do not leave out accountability, transparency, ethics, and trustworthiness of AI…and that is just a start.

A straightforward, simpler approach would be to describe and define AI in overarching terms. AI encompasses both breadth and depth of complex algorithms and specific coding that will eventually allow a silicon-based computing device to be indistinguishable from a carbon-based counterpart when sensing, reasoning, acting autonomously, and adapting to change.

Until that point of being indistinguishable, we can look at several examples of how AI is defined.

- The State of Michigan, House Bill 5143 defines "artificial intelligence" as a machine-based system that can, for a given set of human-defined objectives, make predictions, recommendations, or decisions influencing real or virtual environments.[4]

- The AI Risk Management Framework, developed by the National Institute of Standards and Technology(NIST), approaches defines AI as an engineered or machine-based system that can, for a given set of objectives, generate outputs, such as predictions, recommendations, or decisions influencing real or virtual environments. AI systems are designed to operate with varying levels of autonomy.[5]

Given the reality that there is no one universally accepted definition of AI, Dr. Justin Romberg, the Schlumberger Professor and the Associate Chair for Research in the School of Electrical and Computer Engineering and the Associate Director for the Center for Machine Learning at Georgia Tech, quoted in Joshua Stewart's "What is Artificial Intelligence," provides a succinct and simple definition.... "AI is when a computer or other automated agent makes a decision that a human could make but does so without human input."[6]

In summary, generative AI refers to AI systems designed to generate content, such as text, images, music, or code. Examples include tools like ChatGPT, DALL·E, Microsoft CoPilot, Google Gemini, and Stable Diffusion. It emphasizes the generative capability of the AI.

While general AI (GAI) stands for General Artificial Intelligence or Artificial General Intelligence. It describes a theoretical form of AI capable of performing any intellectual task that a human can do, demonstrating flexible and broad problem-solving abilities. Unlike generative AI, GAI is still a concept and does not yet exist.

THE MAIN TYPES OF AI

AI can be categorized into three main types based on its capabilities:

1. Narrow AI
2. General AI
3. Superintelligent AI

Narrow AI, also known as Weak AI, is designed for specific tasks like facial recognition or language translation and is the only form of AI currently in use. An example of narrow AI is Siri (from Apple). Siri is a virtual assistant that performs specific tasks like setting reminders, answering questions, and controlling smart devices, but it cannot perform tasks outside its programmed domain.

General AI, or Strong AI, would have the ability to understand, learn, and apply knowledge across a wide range of tasks like a human. General AI, capable of performing any intellectual task a human can do across multiple domains, is still a theoretical concept and has not been realized in current technology.

Superintelligent AI would surpass human intelligence in virtually every field, though these latter types remain speculative and theoretical as of today.

TYPES OF AI BY FUNCTION

Reactive AI

The most basic but still quite useful AI is called reactive AI because it reacts to existing conditions as its name suggests. Reactive AI functions the way it was programmed with a predictable output based on the input it receives. These reactive machines will respond to identical situations in the same way every time. Spam filters and the Netflix recommendation engine are other examples of reactive AI.[7]

Limited Memory

Limited memory machines have short-term memory that allows them to temporarily store acquired experiences and act based on them. Just as their name suggests, they can look into the past and create a memory – not the distant past, but still. This type of AI is more advanced than reactive machines. A perfect example of a machine with limited memory is self-driving cars.[8]

Theory of Mind

Theory of mind and self-aware AI are theoretical types that could be built in the future. As such, there are not any real-world examples yet. If it is developed, the theory of mind AI could have the potential to understand the world and how other entities have thoughts and emotions. In turn, this affects how they behave concerning those around them.[9]

Self-Aware AI

Self-Aware AI is a kind of functional AI class for applications that would possess super AI capabilities. If ever achieved, it would have the ability to understand its own internal conditions and traits along with human emotions and thoughts. It would also have its own set of emotions, needs, and beliefs.[10]

GENERATIVE AI: RISKS AND BENEFITS

Generative-AI has the potential to provide significant benefits to organizations.

However, generative AI can create additional risks for organizations. One risk associated with generative AI is exactly what makes generative AI "intelligent," the use of excessively large amounts of data. That data can come from both known (substantiated) and unknown (unsubstantiated) sources. The trustworthiness of data is a critical step in overall AI risk mitigation.

Generative AI also can create harmful content and generate mis- and/or disinformation, what the AI community calls "hallucination." Failing to substantiate content generated by generative AI models may also place the organization at risk, especially if organization personnel rely upon generated responses that are false, misleading, or outright incorrect.

Generative AI has the capability to introduce risk into existing areas such as data and personal privacy (i.e., protected health information), information technology (IT) system security, as well as enterprise policies, and procedures. Thus, raising ethical considerations in the use of unchecked, unsubstantiated generative AI models.

Another risk of using and relying on generative AI, lies with the data itself.

Since many AI systems rely on huge amounts of data to perform well, it is important to understand how data are influencing the behavior of the AI systems. For example, if the training data is biased, that is, it is not balanced or inclusive enough, the AI system trained on such data will not be able to generalize well and will possibly make unfair decisions that can favor some groups over others.[11] Relying on generative AI-generated content based on biased data places additional risk on business operations and perceptions of the organization by its customers.

The reader is encouraged to review chapters in this book that specifically address AI ethics, trustworthiness, risk, accountability, and continuous monitoring.

WHAT IS ARTIFICIAL INTELLIGENCE IN TERMS OF TODAY'S STATE OF TECHNOLOGY?

AI is part of the larger field of data science, the field of study that combines domain expertise, programming skills, and knowledge of mathematics and statistics to extract meaningful insights from data.

As was previously discussed, AI allows a silicon-based unit (i.e., a computer) to solve problems (and increasingly complex ones) that were originally solved by carbon-based life forms (i.e., humans). Basically, AI simulates human intelligence; at least that is the ultimate end game.

Spend any length of time in the field of AI, you will inevitably encounter two subsets of AI, (NLP), and ML.

NLP provides AI with the capability to read, interpret, and understand human language. NLP's focus is on computer–human language interactions, specifically how to program computers to process and analyze large amounts of natural language data. The goal of NLP is for computers to eventually be able to recognize and process text and languages in the same way humans do.

While ML as the name suggests, provides the basis for AI systems to learn and improve responses to queries through experience that does not require specific programming to further an ever-increasing level of intelligence. The goal of ML is to find patterns in very large amounts of data and then to make predictions based on those observed patterns. Machine learning is primarily concerned with accuracy and pattern recognition.

A subset found within ML is DL. DL is a method in AI that teaches computers to process data in a way that is inspired by the human brain. DL models can recognize complex patterns in pictures, text, sounds, and other data to produce accurate insights and predictions.

Finally, a subset of ML and at the heart of DL algorithms we find neural networks, also known as Artificial Neural Networks (ANNs) or Simulated Neural Networks (SNNs). Their name and structure are inspired by the human brain, mimicking the way biological neurons signal to one another. Neural networks rely on training data to learn and improve their accuracy over time.

We use NLP and ML many times throughout the day and many of us never realize it. When was the last time you asked Siri about traffic conditions before you headed out for your morning commute to work or Google Assistant to access information from your calendars and other personal information?

Often the best and most direct approach to explain a complex concept is through the use of a picture. Under the umbrella of data science, Figure 1.2 shows the relationship between AI, NLP, ML, DL, and ANNs. It is important to note, however, that AI is much more complex than shown here since AI includes many other sub-disciplines and techniques, a discussion of which is beyond the scope of this book.

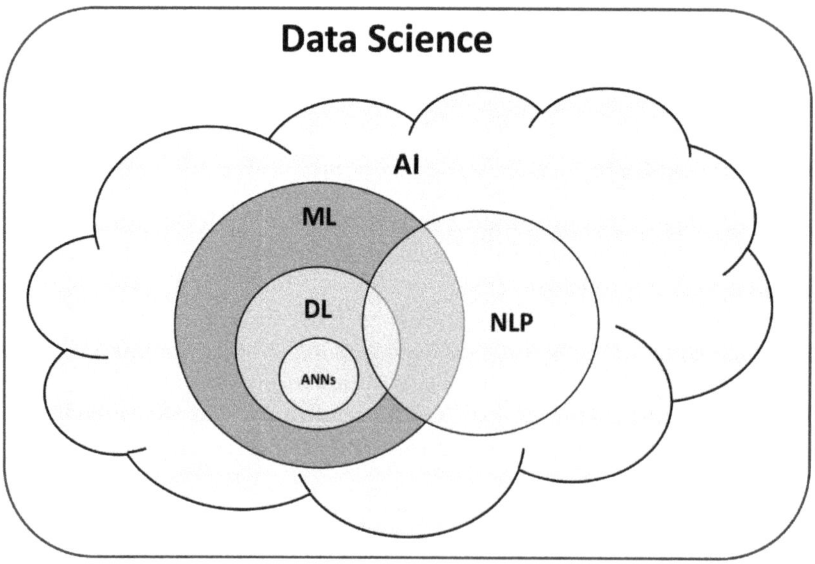

Figure 1.2 The Relationship Between AI, NLP, ML, DL, and ANNs.[12]

TRANSFORMER ARCHITECTURE TECHNOLOGY: WHERE DID CHATGPT COME FROM?

Transformer architecture technology is a DL model architecture introduced in 2017 by Vaswani et al. It has since become a widely adopted and highly influential technology in the field of NLP. Transformer architecture revolutionized the field of NLP and became the basis for many state-of-the-art models, including OpenAI's Generative Pre-trained Transformer (GPT) series.

Before the Transformer Recurrent Neural Networks (RNNs) and their variants, such as Long Short-Term Memory (LSTM) and Gated Recurrent Units (GRU), were widely used for sequence modeling tasks like machine translation and language understanding. RNNs presented limitations in capturing long-range dependencies in sequential data, such as sentences or paragraphs. RNNs process data sequentially, which can be inefficient for long sequences and lead to difficulties in capturing long-term dependencies.

In contrast, the Transformer architecture relies on a self-attention mechanism to capture relationships between different words or positions in a sequence. It allows the model to focus on various parts of the input sequence and determine the importance of each word in the context of the entire sequence. Self-attention enables the model to capture long-range dependencies effectively.

The architecture consists of an encoder-decoder structure. The encoder takes an input sequence and processes it, producing a representation that captures the contextual information of each word. The decoder then uses this representation to generate an output sequence, typically used for tasks such as language translation or text generation.

One key advantage of the Transformer architecture is its parallelizability. Since the model does not rely on sequential processing, computations for different parts of the input sequence can be performed concurrently, making it easier to take advantage of modern hardware, such as GPUs or TPUs, to accelerate training and inference.

Since its introduction, the Transformer architecture has become a cornerstone in NLP research. It has been extended and adapted for various tasks, including language, but not limited to understanding, question-answering, and text generation.

The Transformer architecture has achieved state-of-the-art performance on various NLP tasks, including machine translation, question-answering, text summarization, and sentiment analysis. It has been widely adopted in many NLP applications and forms the basis of models like Bidirectional Encoder Representations from Transformers and GPT. These models have demonstrated remarkable language understanding and generation capabilities, revolutionizing the field of NLP.[13]

To sum up our review of AI and the various subsets that make up the broader field of AI, we are left with one elementary question… "How does AI use data to make informed decisions?"

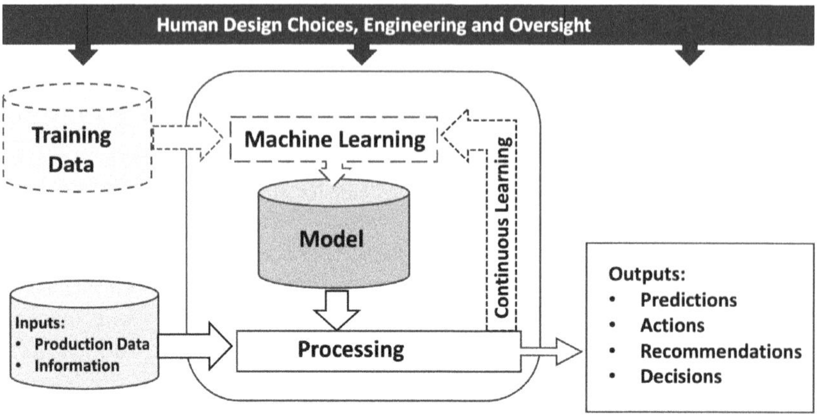

Figure 1.3 AI Functional Overview.[14]

The answer is uncomplicated and straightforward...AI uses data to make informed decisions through a process called "machine learning." ML is a subset of AI that enables systems to learn from data, identify patterns, take actions, and make recommendations, predictions, or decisions based on that data (see Figure 1.3).

> AI is neither built nor deployed in a vacuum, sealed off from societal realities of discrimination or unfair practices. Understanding AI as a socio-technical system acknowledges that the processes used to develop technology are more than their mathematical and computational constructs. A socio-technical approach to AI considers the values and behavior modeled from the datasets, the humans who interact with them and the complex organizational factors that go into their commission, design, development, and ultimate deployment.[15]

The following chapters of this book examine key functions and operations integral to AI systems. The approach to this examination is that of an internal auditor or risk/security assessment professional tasked with auditing/assessing AI systems under development, acquired via third-party or currently deployed, and the organizational environment in which these systems operate.

NOTES

1 The author, to demonstrate the breadth, depth, utility, and application of AI on a limited and controlled basis, utilized GPT-4o, OpenAI's large-scale language-generation model, to provide recommendations to specifically worded technical inquiries. Any generated recommendation provided was fully substantiated and vetted against published technical information and research before the incorporation of any recommended Gen-AI-provided response.

2 The reader should note that any identification and discussion of products and vendors in this work is solely intended to provide informative content and does not constitute an endorsement or promotion of any specific product, brand, or vendor by the author or publisher. The author has received no compensation, financial or otherwise, from any vendor, organization, or third party identified in this work. When acquiring any software product, readers are encouraged to conduct their research and make informed decisions based on their organization's individual needs and preferences.
3 NASA Office of the Inspector General. Figure 1.2 Visualization of the history of AI using select events from the National Institute of Justice (NIJ), September 30, 2018, "A Brief History of Artificial Intelligence," nij.ojp.gov; https://nij.ojp.gov/topics/articles/brief-history-artificial-intelligence, federal guidelines, and NASA missions. Selected AI events occurring after 2022 added by the author.
4 Representative Matt Bierlein, November 8, 2023, "Artificial Intelligence H.B. 5143," Michigan Legislature (.gov), https://legislature.mi.gov/documents/2023–2024/billanalysis/Senate/pdf/2023-SFA-5141-L.pdf.
5 NIST, July 2024, "Artificial Intelligence Risk Management Framework: Generative Artificial Intelligence Profile NIST AI 600–1," Adapted from: OECD Recommendation on AI:2019; ISO/IEC 22989:2022, National Institute of Standards and Technology, www.nist.gov/itl/ai-risk-management-framework.
6 Stewart, J., June 4, 2024, "What is Artificial Intelligence," Helluva Engineer magazine Spring 2024, College of Engineering, Georgia Tech, https://coe.gatech.edu/magazine/2024/spring/what-is-artificial-intelligence.
7 Marr, B., May 2024, "What are the Four Types of AI?" Bernard Marr & Co., https://bernardmarr.com/what-are-the-four-types-of-ai.
8 Sahota, N., September 27, 2022, "7 Types of Artificial Intelligence (With Examples)," Neil Sahota, www.neilsahota.com/7-types-of-artificial-intelligence-with-examples.
9 Coursera Staff, December 19, 2004, "4 Types of AI: Getting to Know Artificial Intelligence," www.coursera.org/articles/types-of-ai.
10 IBM Data and AI Team, October 12, 2023, "Understanding the Different Types of Artificial Intelligence," IBM, www.ibm.com/think/topics/artificial-intelligence-types.
11 High-Level Expert Group on AI (AI HLEG), April 8, 2019, "A Definition of AI: Main Capabilities and Disciplines," https://ec.europa.eu/futurium/en/ai-alliance-consultation.1.html.
12 Marcella, A, 2025, Figure 1.2, "The Relationship Between AI, NLP, ML, DL, and ANNs," drawn by author.
13 OpenAI, January 9, 2024 "What is Transformer Architecture Technology?" GPT-4o, OpenAI's Large-Scale Language-Generation Model.
14 Cotton, P., Patel, M., Wei, W., May 2022, Figure 1.3, "The foundational standards for AI ISO/IEC 22989 and ISO/IEC 23053, ISO/IEC AI Workshop," https://jtc1info.org/wp-content/uploads/2022/06/03_08_Paul_Milan_Wei_The-foundational-standards-for-AI-20220525-ww-mp.pdf
15 Schwartz, R., Vassilev, A., Greene, K., Perine, L., Burt, A., Hall, P., March 2022, "Towards a Standard for Identifying and Managing Bias in Artificial Intelligence," National Institute of Standards and Technology, NIST Special Publication 1270, https://nvlpubs.nist.gov/nistpubs/SpecialPublications/NIST.SP.1270.pdf.

Chapter 2

AI Algorithms

> Algorithm audit can contribute to the trustworthiness of AI technology, improving its impact on the organization's competitiveness. Algorithm audit contributes to making your AI system lawful, ethical, and robust; the three components of a trustworthy AI
>
> – Juan at PickleTech

WHAT ARE AI ALGORITHMS?

AI algorithms are a set of instructions or rules that enable machines to learn, analyze data, and make decisions based on that knowledge. These algorithms can perform tasks that would typically require human intelligence, such as recognizing patterns, understanding natural language, problem-solving, and decision-making.

In any discussion of AI algorithms, it is important to also underscore the value of using the right data in the training of algorithms. At the most basic level, the data needs to be relevant to the issue the algorithm is attempting to solve. The right data should be accurate and free from bias as much as possible.[1]

TYPES OF AI ALGORITHMS

The three main types of AI algorithms are **supervised learning, unsupervised learning,** and **reinforcement learning.** Each of these has unique applications based on their structure and functionality.

Supervised Learning

- Description: Supervised learning algorithms learn from labeled data, meaning they are trained on a dataset that includes both input data and the correct output. They use this data to make predictions or classifications.

DOI: 10.1201/9781003646501-3

- Applications:
  - Image Classification: Identifying objects in images, such as detecting cats vs dogs in photos.
  - Spam Detection: Email-filtering systems that classify emails as spam or not spam.
  - Sentiment Analysis: Analyzing text data to determine sentiment (e.g., positive, negative, neutral) in social media posts or product reviews.

Auditing Supervised Learning AI Algorithms – Audit Questions

1. How does the organization handle potential biases in labeled datasets, and what procedures are in place to detect and mitigate these biases?
2. How is labeled data sourced, and what processes are in place to ensure the accuracy and reliability of labels?
3. What controls are in place to monitor the model's performance over time, especially as new labeled data is added to the system?

Unsupervised Learning

- Description: Unsupervised learning algorithms work with unlabeled data. They identify patterns, groupings, or structures within the data without explicit instructions on what to predict or classify.
- Applications:
  - Customer Segmentation: Dividing customers into groups based on purchasing behavior to improve targeted marketing.
  - Anomaly Detection: Detecting unusual patterns in network traffic to identify potential security threats.
  - Topic Modeling: Discovering hidden topics in text data, such as grouping news articles by topic.

Auditing Unsupervised Learning AI Algorithms – Audit Questions

1. How does the organization address potential issues when unsupervised models generate unexpected or undesirable groupings?
2. What methods are used to validate the patterns or groupings identified by unsupervised learning models?
3. What safeguards are implemented to prevent misinterpretation of patterns identified by unsupervised models in critical decision-making processes?

Reinforcement Learning

- Description: Reinforcement learning algorithms learn by interacting with an environment, receiving rewards or penalties based on their actions, and aiming to maximize cumulative rewards.

- Applications:
  - Robotics: Training robots to perform tasks like navigation or object manipulation.
  - Game AI: Developing game-playing agents that learn strategies in games like chess, Go, or video games.
  - Autonomous Vehicles: Enabling self-driving cars to learn how to navigate and make decisions based on traffic rules and road conditions.

Auditing Reinforcement Learning AI Algorithms – Audit Questions

1. How does the organization define and manage the rewards and penalties that guide the reinforcement learning model's behavior?
2. How is the performance of reinforcement learning models monitored in real-world environments, and what contingencies are in place if they perform unpredictably?
3. What measures are in place to ensure that reinforcement learning models do not learn unintended behaviors that could lead to operational risks or ethical concerns?

These types of algorithms provide the foundation for various AI applications, with each type suited to particular problem types and data availability.

TYPES OF HYBRID AI ALGORITHMS

Hybrid approaches in AI combine different types of algorithms, often integrating supervised, unsupervised, and reinforcement learning techniques to take advantage of each method's strengths. These approaches allow for more versatile and adaptable AI systems.

The following are common types of hybrid AI approaches and examples of their applications:

Semi-Supervised Learning

- Description: Semi-supervised learning combines supervised and unsupervised learning. It leverages a small amount of labeled data along with a large amount of unlabeled data. This approach is useful when labeling data is expensive or time-consuming.
- Applications:
  - Medical Imaging: In medical imaging analysis, obtaining labeled data (e.g., images with disease annotations) can be challenging. Semi-supervised learning helps identify abnormalities using a few labeled scans and many unlabeled ones.
  - Speech Recognition: In speech-to-text systems, where labeled audio data is scarce, semi-supervised models can learn to transcribe speech with minimal labeled data, improving accuracy over time with new data.

- Sentiment Analysis: It is applied to sentiment analysis tasks in languages with limited labeled data. The model learns from a small, labeled dataset and refines its performance with larger amounts of unlabeled text.

Auditing Semi-Supervised Learning AI Algorithms – Audit Questions

1. How does the organization address potential biases in both labeled and unlabeled datasets in semi-supervised learning?
2. How is the effectiveness of semi-supervised learning models monitored over time, especially as additional unlabeled data is incorporated?
3. What measures are taken to validate the patterns or insights derived from the predominantly unlabeled data?
4. What processes are in place to ensure the accuracy and reliability of the small amount of labeled data used in semi-supervised learning models?
5. What safeguards exist to prevent misuse or misinterpretation of semi-supervised learning model outputs in critical applications, such as medical imaging?

Self-Supervised Learning

- Description: Self-supervised learning creates its own labels from data without human annotation. The model learns to predict parts of data from other parts, which helps it understand structures and patterns without manual labeling.
- Applications:
  - Natural Language Processing: Many language models, like BERT and GPT, use self-supervised learning to predict missing words in sentences or generate responses based on context. This approach has revolutionized applications like machine translation, chatbots, and summarization.
  - Computer Vision: In tasks like object detection, self-supervised models can predict spatial structures and identify relationships within an image. This is useful for applications such as autonomous driving, where labeled visual data is difficult to obtain.
  - Recommender Systems: Self-supervised learning is used to predict user preferences by learning from user behavior without explicit labels. Platforms like e-commerce or streaming services utilize it to make personalized recommendations.

Auditing Self-Supervised Learning AI Algorithms – Audit Questions

1. How does the organization address potential overfitting to specific data patterns, given the model's reliance on its own generated labels?
2. How does the organization ensure that the labels generated by the model itself are accurate and meaningful for self-supervised learning applications?

3. How is model performance validated in scenarios where human-labeled data is unavailable or impractical to obtain?
4. What mechanisms are in place to detect and address any unintended biases introduced by self-supervised learning models?
5. What processes are followed to ensure that self-supervised learning models remain effective and reliable over time?

Multi-Agent Reinforcement Learning (MARL)

- Description: MARL uses multiple reinforcement learning agents working together or in competition within a shared environment. Agents learn optimal behaviors based on both their actions and the actions of other agents, making this ideal for complex systems.
- Applications:
  - Traffic Management: In smart city traffic systems, MARL coordinates traffic lights by having each traffic light as an agent that learns to optimize flow based on neighboring lights and traffic patterns.
  - Team-Based Games: In multiplayer video games, MARL allows different agents (representing players or game entities) to cooperate or compete, leading to more realistic and challenging opponents.
  - Resource Allocation: In network optimization, such as wireless communication networks, MARL agents optimize resource allocation by learning how to balance network loads or allocate bandwidth efficiently.

Auditing Multi-Agent Reinforcement Learning AI Algorithms – Audit Questions

1. How are reward and penalty structures designed to avoid encouraging unintended behaviors among MARL agents?
2. How does the organization evaluate and address potential risks from the dynamic and potentially competitive interactions between agents?
3. How does the organization manage and monitor the interactions between multiple reinforcement learning agents to prevent conflicts or suboptimal outcomes?
4. What measures are taken to ensure that each agent within MARL models acts in alignment with the organization's overall goals?
5. What monitoring processes are in place to ensure MARL agents' behavior remains consistent with evolving organizational objectives or regulatory requirements?

Model-Based Reinforcement Learning (MBRL)

- Description: Model-based reinforcement learning combines traditional reinforcement learning with a predictive model of the environment. Instead of directly learning through trial and error in the real environment, the model simulates future states based on actions.

This approach speeds up learning and reduces the need for extensive interaction with the real world, which is useful when real-world experimentation is costly or risky.
- Applications:
  - Robotics: MBRL allows robots to simulate actions in a virtual environment, which reduces wear and tear and speeds up training. For example, a robot arm can practice grasping objects virtually before testing the learned skills in the real world.
  - Autonomous Vehicles: Self-driving cars use MBRL to simulate driving scenarios and learn optimal responses to events like sudden obstacles or pedestrian crossings. This simulation-driven approach allows cars to improve without actual road testing for every scenario.
  - Healthcare: In personalized treatment planning, MBRL helps simulate the effects of various treatment options on virtual patient models. This allows for safe testing of treatment responses and adjustments before applying them in real life.

Auditing Multi-Based Reinforcement Learning AI Algorithms – Audit Questions

1. How are actions simulated by MBRL models tested for safety and feasibility before they are applied in real-world scenarios?
2. How does the organization handle situations where the simulated environment fails to accurately represent real-world complexities?
3. How is the accuracy and reliability of the predictive model used in MBRL validated, especially as it simulates future states?
4. What measures are taken to ensure that MBRL models do not rely on inaccurate predictions, particularly in high-stakes applications like autonomous vehicles?
5. What processes are in place to ensure the model is updated with real-world data to maintain accuracy over time?

Generative Adversarial Networks (GANs)

- Description: GANs consist of two neural networks – the generator and the discriminator – that compete in a process called adversarial training. The generator creates fake data, while the discriminator tries to distinguish between real and fake data. This interaction helps the generator produce increasingly realistic outputs, enabling the creation of high-quality synthetic data.
- Applications:
  - Image and Video Synthesis: GANs are widely used to create realistic images, videos, or artwork. Applications include generating human faces, realistic product images for e-commerce, and creating photorealistic avatars for virtual reality.

- Medical Data Augmentation: GANs generate synthetic medical images, such as MRIs or X-rays, to expand small datasets. This helps improve diagnostic models by providing more diverse training samples without additional patient data.
- Data Privacy and Anonymization: GANs create realistic but synthetic datasets that mimic real data distributions, allowing companies to use data without risking privacy. This approach is popular in industries like finance and healthcare, where data sensitivity is high.

Auditing Generative Adversarial Networks AI Algorithms – Audit Questions

1. How are GAN-generated outputs verified to ensure they align with the intended application's requirements, especially in areas like medical data augmentation?
2. How does the organization monitor and evaluate the quality and realism of data generated by GANs to avoid potential misuse?
3. What controls are implemented to prevent the discriminator in GANs from becoming overly optimized, leading to reduced generator diversity?
4. What procedures are in place to ensure the ethical and legal use of synthetic data generated by GANs, particularly in sensitive areas like data privacy?
5. What processes are in place to detect and address any biases introduced in synthetic data generation, particularly if used in training other models?

AUDITING AI ALGORITHMS

What Is an Algorithm Audit?

In an algorithmic audit, an independent party may evaluate an algorithmic system for bias, but also accuracy, robustness, interpretability, privacy characteristics, and other unintended consequences. The audit would identify problems and suggest improvements or alternatives to the developers. Beyond improving systems, algorithmic auditing can also help to build consumer and client trust if its results are made public.[2]

As defined by Metaxa et al., an algorithm audit is a method of repeatedly and systematically querying an algorithm with inputs and observing the corresponding outputs in order to draw inferences about its opaque inner workings.

What all algorithm audits do have in common is an aim to test whether some deficiency (discrimination, bias, or something else) exists in an algorithmic system or not without direct access to the internals of that system.[3]

Engler, writing in "Auditing Employment Algorithms For Discrimination," identifies 10 important criteria that help to evaluate the quality of algorithm audits performed internally or via an external third party.

1. Auditor independence: An algorithmic audit should consciously structure contracts to maintain the independence of the auditor and subsequently the report of the auditor's findings.
2. Representative analysis: An algorithmic audit does not need to look at every deployed model, but a representative sample of models needs to be examined to be able to evaluate the controls that may (or may not) be present. A random, stratified sample may prove prudent whereby the auditor might take random samples of different types of models.
3. Data collection process: An algorithmic audit should actively consider the data collection and cleaning process, especially including any exclusion of observations, top-coding of outliers, consolidation of categorical variables, or missing data imputation and their possible effects on model training and outcomes.
4. Training data representativeness: Datasets used for training machine-learning models should be evaluated for representativeness to the extent that demographic data is available.
5. Dependency analysis: An algorithmic audit should consider the effects of, and report on, the software packages and libraries that the audited company's algorithmic system builds on.
6. Documentation review: An algorithmic audit should review the documentation for models and ensure that the documentation accurately communicates the functionality of the models.
7. Performance on historical data: An algorithmic audit should also consider how trained models performed on the data used to develop the model.
8. Hypothetical data: Auditors can generate synthetic data to stress test the model-building process. This process could check whether automated or human-in-the-loop oversight would catch a problematically trained model.
9. Problem definition: An algorithmic audit should carefully consider the outcome variables chosen and how that might enable biases.
10. Frequency: The algorithmic audit should consider how models update over time and what that entails for model drift, especially the possibility of performance degradation. This consideration should lead to a recommendation on how often a modeling audit would need to occur.[4]

For Additional Audit Questions on Auditing AI Algorithms, see Appendix 2.

AUDITING AI ALGORITHMS – SUPPLEMENTAL MATERIAL

The auditor, security, and risk professional should review and incorporate concepts from the following documents into the audit of AI algorithms:

IEEE 7003 Draft Standard for Algorithmic Bias Considerations

> This draft standard describes processes and methodologies to help users address issues of bias in creating algorithms. Possible elements include (but are not limited to): criteria for the selection of validation data sets for bias quality control; guidelines on establishing and communicating the application boundaries for which the algorithm has been designed and validated to guard against unintended consequences arising from out-of-bound application of algorithms; suggestions for user expectation management to help mitigate bias due to incorrect interpretation of systems outputs by users (e.g. correlation vs. causation).
>
> *IEEE 7003 Standard for Algorithmic Bias Considerations [https://standards.ieee.org/ieee/7003/11357/]*

NOTES

1 Kerner, S., Tabsharani, F., October 16, 2024, "Types of AI algorithms and how they work," TechTarget, www.techtarget.com/searchenterpriseai/tip/Types-of-AI-algorithms-and-how-they-work.
2 Engler, A., March 12, 2021, "Auditing employment algorithms for discrimination," The Brookings Institution, www.brookings.edu/articles/auditing-employment-algorithms-for-discrimination.
3 Metaxa, D., Park, JS, Robertson, R., et al., 2021, "Auditing Algorithms", Foundations and Trends® in Human-Computer Interaction: Vol. 14, No. 4, pp 272–344. DOI: 10.1561/1100000083, https://hci.stanford.edu/publications/2021/FnT_AuditingAlgorithms.pdf.
4 Engler, A., March 12, 2021, "Auditing employment algorithms for discrimination," The Brookings Institution, www.brookings.edu/articles/auditing-employment-algorithms-for-discrimination.

REFERENCES

"Auditing The Quality Of Datasets Used In Algorithmic Decision-Making Systems," European Parliamentary Research Service, Scientific Foresight Unit (STOA) PE 729.541, July 2022, ISBN: 978-92-846-9681-9, doi: 10.2861/98930, QA-08-22-267-EN-N, EU Auditing Datasets Used in Algorithmic Decision-Making [www.europarl.europa.eu/RegData/etudes/STUD/2022/729541/EPRS_STU(2022)729541_EN.pdf]

Marian, Iurian., and Groza, Adrian. (2022), "Towards Assuring Conformance of AI Systems,"

Assuring Conformance of AII Systems [www.researchgate.net/publication/360709226_Towards_assuring_conformance_of_AI_systems]

European Commission, Content Directorate-General for Communications Networks, and Technology. 2019. Ethics Guidelines For Trustworthy AI. Publications Office. EU Ethics Guidelines for Trustworthy AI [https://ec.europa.eu/newsroom/dae/document.cfm?doc_id=60419]

Lloret, J., 21 February 2020, "WHITE PAPER On Artificial Intelligence – A European Approach to Excellence and Trust," AI - A European Approach to Excellence and Trust [https://ec.europa.eu/futurium/en/system/files/ged/white_paper_ai_19_02_2020.pdf]

European Commission, Directorate-General for Communications Networks, Content and Technology, (2020). The Assessment List for Trustworthy Artificial Intelligence (ALTAI) for self-assessment, Publications Office. Assessment list for Trustworthy AI [https://data.europa.eu/doi/10.2759/791819]

Raji, I. D., Smart, A., White, R. N., Mitchell, M., Gebru, T., Hutchinson, B., and Barnes, P. (2020, January). "Closing the AI Accountability Gap: Defining an End-to-End Framework for Internal Algorithmic Auditing." In *Proceedings of the 2020 Conference on Fairness, Accountability, and Transparency*, Closing the AI Accountability Gap 33–44 pages. [https://arxiv.org/abs/2001.00973]

Benbouzid, D., et al, 21 May 2024, "Pragmatic Auditing: A Pilot-Driven Approach For Auditing Machine Learning Systems," arXiv:2405.13191v1 [cs.LG], Auditing Machine Learning Systems [https://arxiv.org/html/2405.13191v1#bib.bib24]

Chapter 3

Key Metrics for Generative AI

Key metrics for evaluating generative AI tools may be divided into separate yet related categories:

Performance of the Model is used to assess or evaluate the tool's response time, precision, and accuracy.

Value Delivered assists in assessing the tool's ability to reduce the time and effort spent to perform tasks, increase throughput, and increase demand generation.

Operational areas within the organization in which metrics useful for assessing generative AI include:

Measuring ROI to determine if and how a machine learning (ML) program or project delivers meaningful value.

The following Key performance indicators (KPIs) provide a comprehensive view of an ML project's impact, allowing stakeholders to gauge its alignment with the desired ROI of delivering meaningful value. Auditors should assess whether these or other KPIs are utilized by the organization under review to measure generative AI programs or projects' ROI.

1. Cost Savings: Measures the reduction in operational expenses due to the ML solution.
   Example: A predictive maintenance model reduces downtime costs by 20%.
2. Revenue Growth: Tracks the increase in revenue directly attributed to the ML model's impact.
   Example: A recommendation system boosts sales by 15% through personalized product suggestions.
3. Model Accuracy: Measures the percentage of correct predictions, crucial for assessing performance in predictive tasks.
   Example: A fraud detection model achieves 98% accuracy in identifying fraudulent transactions.
4. Time Savings: Evaluates the reduction in time required for processes automated or assisted by ML.
   Example: An automated data entry model decreases processing time by 40% for customer service requests.

DOI: 10.1201/9781003646501-4

5. Customer Satisfaction (CSAT): Monitors customer satisfaction levels with interactions influenced by the ML system.
   Example: After implementing an AI-powered chatbot, CSAT scores improve by 10%.
6. Error Rate Reduction: Measures the decrease in errors within processes optimized by ML.
   Example: An image recognition model reduces diagnostic errors in medical imaging by 15%.
7. Operational Efficiency: Assesses improvements in overall process efficiency due to ML.
   Example: A supply chain optimization model reduces inventory carrying costs by 25%.
8. Customer Retention Rate: Measures the percentage of customers retained due to improvements brought by the ML model.
   Example: A churn prediction model helps increase customer retention by 8%.
9. Adoption Rate: Tracks the usage and integration of the ML model within workflows or systems.
   Example: 85% of customer service representatives actively use the sentiment analysis tool for support calls.
10. Scalability: Evaluates the ML model's performance under increased workloads, ensuring it can grow with demand.
    Example: The model maintains performance levels and latency even when handling a 50% increase in daily transactions.[1]

Measuring goal completions, to identify how many tangible and desirable business outcomes a model achieves.

Fidelity gauges the success of generative AI-based systems by assessing the similarity between generated output and real data. A high-fidelity score indicates the model's proficiency in producing realistic and accurate results, which is essential for building trust in the technology/model.

Task performance measures how well the model responds to a given prompt, such as summarizing some text, solving a math problem, or performing common-sense reasoning. Relevant metrics include the following:

- A benchmark used to evaluate a model's task performance capabilities is the **Massive Multitask Language Understanding (MMLU)**, which is designed to evaluate a language model's proficiency across a wide range of subjects and tasks. It measures the model's performance on a diverse set of questions, covering various disciplines from the humanities to the sciences, using multiple-choice format questions at different difficulty levels. MMLU assesses a model's breadth of knowledge and ability to handle tasks across multiple domains, reflecting its general language understanding capabilities.

- **Generation consistency**, which measures whether similar prompts in the same context can lead to almost semantically similar responses.
- **Prompt sensitivity**, which measures how detailed a prompt needs to be to get the optimal response from the LLM.

Safety metrics help test for risks such as ethical concerns about generative AI, truthfulness, toxicity, and security. This can include measuring the prevalence of racially biased responses, AI hallucinations, or leaks of confidential information.

Personality metrics related to personality dimensions or the emotional intelligence of the generative AI models that power chatbots to help better understand and compare their behavior.

Accuracy measures how well a model's predictions or outputs align with the desired results. Assessing accuracy is often easier by domains, such as:

- Perplexity is a metric that evaluates a language model's ability to predict the next word in a word sequence.
- Inception score (IS) is a mathematical algorithm that measures the quality of generative AI images.
- Fréchet inception distance (FID) analyzes images generated by generative adversarial networks for realism and diversity.
- Precision is a metric that measures the number of correct predictions made by a generative AI model.
- Recall details the ratio of positive samples classified by a model as positive vs the total number of positive samples generated.
- F1 score measures an AI model's accuracy using the precision and recall factors.
- Bilingual evaluation understudy (BLEU) is a metric used to automatically evaluate machine-translated text against reference text.
- Recall-oriented understudy for gisting evaluation (ROUGE) measures the quality of a machine-translated summary against one created by a human.
- Metric for evaluation of translation with explicit ordering (METEOR) scores machine translations based on word-to-word matching against a reference translation.
- Consensus-based image description evaluation (CIDEr), measures a machine-generated sentence against human-generated information that is known to be real and true.
- Manual evaluation is when a human compares machine-generated results on a case-by-case basis.

Inference speed quantifies the model's performance in terms of speed and efficiency at runtime. The latency of the model is typically measured in iterations per second, which directly affects the inference cost of the model.

Lower latency leads to reduced compute cost, a smaller carbon footprint, and an improved overall user experience.[2]

AUDITING GENERATIVE AI METRICS – AUDIT QUESTIONS

The previous section introduced generative AI metrics and related benchmarks that may be used to assess the functionality, effectiveness, and efficiency of generative AI tools. These benchmarks and metrics serve as the foundation for auditing how the organization assesses and evaluates its investment, use, and expectation of results from enterprise-wide, implemented AI solutions.

Auditors, security, and risk professionals should determine which metrics the organization employs in its AI evaluation, acquisition process, and testing processes, the enterprise-wide risk management program, and overall AI security protocols.

The following is a representative list of audit questions, that focus on the organization's use of metrics.

Measuring ROI

1. What measurable business outcomes or KPIs were defined at the outset of the ML project, and how well does the project align with these objectives?
2. How has the ML program impacted efficiency, revenue generation, cost savings, or other value-driven metrics compared with the projected or baseline metrics?
3. What indirect or non-monetary benefits (e.g., improved customer satisfaction, faster decision-making, or enhanced process accuracy) has the ML project provided, and how are these benefits quantified or assessed?
4. What is the time frame over which ROI is being measured, and are the observed outcomes sufficient to justify the resources invested (such as time, budget, and human capital) in the project?
5. How does the project manage and mitigate risks or unforeseen costs that could affect the overall ROI, and are there contingency plans in place to maintain or improve the return on investment?

Measuring Goal Completions

1. What specific business goals was the model designed to achieve, and how are these goals aligned with the organization's strategic objectives?
2. How accurately and consistently does the model perform actions that lead to goal completion, and what factors contribute to its success or failure in achieving these goals? How is this substantiated?

3. What proportion of the model's outputs directly contributes to tangible outcomes, and how are these outcomes measured and validated over time?
4. Are there any patterns in the types of goals the model is more successful in achieving, and what insights can be drawn to enhance goal completion rates in other areas?
5. How does the model's performance in goal completions compare with previous benchmarks or alternative approaches (human-independent processes), and what improvements have been made based on past performance?

Fidelity

1. How is fidelity measured in terms of similarity, and what specific benchmarks or criteria are used to assess alignment with real data?
2. What methods are in place to quantify or qualify the similarity between generated outputs and real data, and how reliable are these methods in detecting subtle discrepancies?
3. To what extent does the generated output retain essential characteristics of real data, such as style, structure, and context, without introducing artifacts or distortions?
4. How does the fidelity of generated outputs vary across different types of data, and what factors (e.g., data complexity, diversity) influence this variation?
5. What processes are used to periodically review and update fidelity standards, ensuring that generated outputs continue to meet evolving real-data expectations?

Task Performance

Massive Multitask Language Understanding (MMLU)

1. How effectively does the model understand and interpret the nuances of each prompt across different subject areas and complexity levels?
2. What specific challenges or limitations does the model demonstrate when attempting to perform tasks that require common-sense reasoning, and how does it overcome them?
3. How consistent are the model's responses when presented with similar prompts in different contexts or formats?
4. In cases where the model fails to provide an accurate response, what are the identifiable factors (e.g., prompt ambiguity, topic complexity) that might contribute to this outcome?
5. How does the model handle prompts that require integrating knowledge from multiple domains, and does it accurately synthesize information when required?

Generation Consistency

1. How consistently does the model generate semantically similar responses when presented with similar prompts in the same context, and what level of variation is considered acceptable?
2. What methods are used to evaluate the semantic similarity between responses, and how are these methods calibrated to ensure they capture subtle differences that might impact consistency?
3. How does the model handle variations in language or phrasing within similar prompts, and does it maintain the intended meaning across different prompt formulations?
4. Are there specific contexts or types of prompts where the model struggles to maintain consistency, and what patterns or factors contribute to these inconsistencies?
5. What steps are in place to monitor and improve generation consistency over time, and how is feedback on consistency issues incorporated into model updates?

Prompt Sensitivity

1. How does the level of detail in a prompt affect the quality and accuracy of the responses, and what specific prompt characteristics lead to optimal outcomes?
2. What methods are used to determine the minimal level of detail required in a prompt to elicit accurate responses, and how consistently do these methods apply across various topics and contexts?
3. How does the model respond to prompts with different levels of ambiguity, and does it require additional prompt detail to avoid misunderstandings or misinterpretations?
4. What variations in response quality are observed when prompts range from minimal to highly detailed, and are there diminishing returns in response quality with increased prompt specificity?
5. How are adjustments made to the model to improve prompt sensitivity, ensuring that it provides reliable responses even when prompts lack extensive detail?

Safety

1. What procedures are in place to identify and measure the prevalence of biased or ethically concerning responses, such as those involving racial, gender, or cultural biases, and how are these findings addressed?
2. How is the model evaluated for truthfulness, particularly in areas where misinformation or AI hallucinations could pose risks, and what corrective measures are implemented for inaccuracies?

3. What mechanisms are employed to detect and mitigate the generation of toxic or harmful language, and how is the model updated to mitigate such risks over time?
4. How does the model ensure the confidentiality of sensitive or private information, and what safeguards are in place to prevent the leakage of confidential data in responses?
5. How frequently is the model audited for adherence to safety and ethical standards, and how is feedback from these audits used to refine and enhance the model's safety profile?

Personality

1. How consistently does the model demonstrate key personality dimensions (e.g., openness, agreeableness, conscientiousness) in interactions, and are these traits aligned with the intended user experience?
2. How does the model adapt its responses to exhibit emotional intelligence, such as responding empathetically or appropriately modulating tone based on user emotions or context? How is this substantiated?
3. What mechanisms are in place to assess and maintain personality stability in the model, ensuring it behaves predictably across different conversations and user inputs? What documentation is available to substantiate these mechanisms?
4. How effectively does the model recognize and respond to emotional cues from users, and what improvements are made to enhance its emotional intelligence over time?
5. How is the model's personality calibrated to avoid unintended biases or inappropriate behaviors, and what guidelines are established to ensure its behavior aligns with ethical standards and user expectations? How is this substantiated?

Accuracy

1. What criteria are used to define and measure accuracy for the model, and how well do these criteria align with the specific goals and desired outcomes of the project?
2. How frequently does the model's output deviate from the expected results, and what patterns or factors are observed in cases where accuracy falls short?
3. What methods are employed to validate the model's accuracy, and how rigorous or representative are these methods across different data subsets or scenarios?
4. How does the model's accuracy vary across different input conditions, and are there specific types of inputs where the model consistently underperforms or excels?

5. What steps are taken to improve and maintain accuracy over time, and how are updates or retraining efforts prioritized based on observed accuracy trends?

Auditing Accuracy by Domain

Perplexity

1. How is perplexity being calculated for the model, and what baseline perplexity values are considered acceptable for the specific use case or language domain?
2. What factors influence the model's perplexity across different types of text inputs, and how does the model handle complex or less common language structures?
3. How does the model's perplexity vary with different levels of context provided, and does it perform better with certain types or lengths of input sequences?
4. What specific challenges or patterns are observed in cases where the model has high perplexity, and what steps are taken to reduce it in these instances?
5. How is perplexity used alongside other metrics to evaluate overall model performance, and how are decisions made to balance perplexity with other language generation qualities, such as coherence or creativity? What documentation exists to substantiate such a comparison?

Inception Score (IS)

1. How well does the Inception Score correlate with the perceived quality and diversity of the generated images, and are there specific cases where the IS may not fully capture these aspects?
2. What steps are taken to ensure that the Inception Score is a reliable indicator for the intended use of the images, especially in cases where human judgment of quality may differ? How is this substantiated?
3. How does the Inception Score vary across different categories or styles of generated images, and what patterns are observed in cases of high versus low IS? What different categories, styles, or patterns are examined to ensure low variance?
4. What are the limitations of relying solely on the Inception Score for image quality assessment, and how is it supplemented with other metrics or evaluations?
5. How are improvements to the model's image generation capabilities reflected in changes to the Inception Score, and what benchmarks are set for determining satisfactory image quality?

Fréchet Inception Distance (FID)[3]

1. How effectively does the FID metric reflect both the realism and diversity of generated images and what specific thresholds are used to determine acceptable FID scores for the application?
2. What patterns or trends are observed in FID scores when comparing images across different categories or complexity levels, and how do these insights guide model improvements?
3. How does the model handle variations in input data distribution, and does this impact the FID scores in a way that might indicate potential biases or limitations in image generation?
4. What complementary metrics or evaluations are used alongside FID to ensure a comprehensive assessment of image quality, especially in cases where FID alone may not capture certain aesthetic or contextual aspects?
5. How frequently is FID monitored during model training and iteration, and what strategies are implemented to reduce FID scores as part of the model improvement process? How is this process substantiated?

Precision

1. What criteria define a "correct" prediction in the model's context, and how are these criteria aligned with the specific goals of the model's deployment?
2. How consistently does the model achieve high precision across different types of input data, and are there specific scenarios where precision tends to decrease? What documentation is used to record this consistency (or lack thereof)?
3. What methods are used to validate the precision of the model's predictions, and how robust are these methods against variations in input and context?
4. How does the model's precision compare with other relevant metrics, such as recall, and what trade-offs are considered to balance precision with overall model performance?
5. What strategies are in place to monitor and improve precision over time, and how is feedback on incorrect predictions incorporated into model updates?

Recall

1. How is recall being calculated for the model, and what constitutes a "positive" sample within the context of the model's objectives? Is documentation available that specifically and with detail, shows how recall is being calculated? If so, who maintains this documentation?

2. In what situations does the model struggle to correctly identify positive samples, and what factors contribute to these recall limitations?
3. How does the recall performance vary across different types or categories of data, and are there specific areas where recall needs improvement?
4. What methods are used to validate and ensure the accuracy of recall measurements, and how frequently are these measurements reviewed to detect potential changes over time?
5. How does the model's recall rate compare with precision, and what trade-offs are made between recall and precision to achieve balanced performance?

F1 Score[4]

1. How does the F1 score reflect the balance between precision and recall for this model, and what trade-offs are considered to achieve an optimal balance for the intended application? Who authorizes implementing trade-offs if or when required? How is this authorization substantiated?
2. In what scenarios does the model achieve high or low F1 scores, and what factors contribute to variations in performance across different data subsets? What are the operational benefits/impacts resulting from a high F1 score? What are the operational benefits/impacts resulting from a low F1 score?
3. How are precision and recall individually analyzed and addressed to improve the F1 score, and what specific improvements are prioritized based on the model's objectives?
4. What methods are used to validate the F1 score calculations, and how frequently are these calculations reviewed to detect any potential shifts in model performance? What documentation exists that substantiates the validation of F1 scores is performed?
5. How is the F1 score used in conjunction with other performance metrics to provide a holistic assessment of the model's effectiveness, and what actions are taken if the F1 score does not meet expected benchmarks?

Bilingual Evaluation Understudy (BLEU)

1. How accurately does the BLEU score reflect the quality of the machine translation in terms of fluency, relevance, and readability compared with the reference text?
2. In what types of text or language pairs does the model achieve high BLEU scores, and are there specific patterns where BLEU scores might not fully capture translation quality?

3. How is the BLEU metric weighted for different n-gram levels,[5] and does this weighting appropriately reflect the linguistic nuances required for accurate translation?
4. What additional metrics or human evaluations are used alongside BLEU to ensure a comprehensive assessment of translation quality, particularly in cases where BLEU may not capture subtle context or idiomatic expressions?
5. How frequently are BLEU scores reviewed and recalibrated based on changes in model performance, and what improvements are made to address areas where the BLEU score indicates potential weaknesses in translation quality? What documentation exists that substantiates that a review of BLEU scores is performed? If required, what documentation exists to substantiate that recalibration based on changes in model performance is performed?

Recall-Oriented Understudy for Gisting[6] Evaluation (ROUGE)

1. How well does the ROUGE metric capture the relevance and completeness of key information in the machine-generated summary compared with the human-created reference summary?
2. What types of ROUGE scores (e.g., ROUGE-N,[7] ROUGE-L[8]) are used to evaluate the summary, and how do these variations affect the interpretation of the model's performance?
3. In what ways does the ROUGE metric handle paraphrasing and synonyms, and are there limitations where ROUGE may fail to recognize semantically equivalent expressions?
4. How is ROUGE complemented by other evaluation methods to provide a holistic assessment of summary quality, especially in cases where ROUGE alone may overlook nuanced content differences?
5. How frequently are ROUGE scores reviewed and recalibrated based on changes in model performance or summary requirements, and what actions are taken to improve the model's summarization quality based on ROUGE findings? How is this review substantiated? How is the recalibration, if required, documented? Who is responsible for the review process? Who is responsible for providing sign-off on these processes?

Metric for Evaluation of Translation with Explicit Ordering (METEOR)

1. How well does the METEOR metric capture both exact and approximate word-to-word matches, and how does it account for synonyms, stemming, and paraphrasing in translations?
2. In what ways does METEOR address the importance of word order and sentence structure, and how does this impact its evaluation of translation fluency and coherence?

3. How is the METEOR score calibrated to reflect the relative importance of precision, recall, and alignment in translation quality, and what adjustments are made for different language pairs? How is this calibration documented?
4. What limitations exist in using METEOR as the sole measure of translation quality, and how is it supplemented with other metrics or human evaluations for a more comprehensive assessment?
5. How does METEOR score variation correlate with user feedback or human evaluations, and what insights can be drawn to improve model performance based on METEOR scores?

Consensus-Based Image Description Evaluation (CIDEr)

1. How well does the CIDEr metric reflect the alignment between machine-generated descriptions and human-generated descriptions in terms of relevance, accuracy, and completeness?
2. What specific weighting or scoring mechanisms are used within CIDEr to evaluate the similarity of key phrases, and how does this impact the interpretation of the model's descriptive accuracy?
3. How does CIDEr handle variations in language or synonyms across human-generated references, and does it appropriately account for equivalent expressions in evaluating description quality?
4. In what scenarios does CIDEr fail to capture nuanced or context-specific details in image descriptions, and what complementary metrics are used to address these limitations?
5. How frequently are CIDEr scores reviewed and updated based on feedback or new human references, and what improvements are made to enhance the model's descriptive accuracy based on CIDEr findings?

Manual Evaluation

1. What criteria are established for human evaluators to assess the quality of machine-generated results, and how consistent are these criteria applied across different cases? How are these criteria documented?
2. How are evaluators trained to minimize bias and subjectivity in their assessments, and what guidelines are in place to ensure reliable and replicable evaluations? What documentation exists that describes the training provided to evaluators? Who has developed this training material? What documentation exists that substantiates that the training developers are qualified to produce this type of training material?
3. In what ways are human evaluations compared or calibrated against automated metrics, and how is alignment between manual and automated assessments ensured? How is this comparison substantiated/documented?

4. How are disagreements or inconsistencies in human evaluations resolved, and what processes are in place to address cases where evaluators have differing interpretations?
5. How frequently are manual evaluation processes reviewed and updated, and what feedback mechanisms are used to improve the effectiveness and accuracy of human evaluations over time?

Inference Speed

1. How is inference speed measured under different operating conditions, and what benchmarks are set for acceptable speed and efficiency based on the application requirements?
2. What factors (such as model size, input data complexity, or hardware configurations) impact inference speed, and how are these factors optimized to maintain consistent performance?
3. How does the model's inference speed compare across various deployment environments, and are there specific scenarios where performance significantly differs?
4. What trade-offs are considered between inference speed and accuracy, and how are these trade-offs managed to ensure that speed improvements do not compromise result quality?
5. How frequently is inference speed monitored and tested post-deployment, and what adjustments are made to improve speed as new data or performance insights become available?

NOTES

1 Marcella, A., November 2024, "Machine learning (ML) Program or Project KPIs and Examples," developed by the author to enhance, supplement, and provide context to the concept of using KPIs in measuring gen-AI models, ROI.
2 Source material for this introduction on Gen-AI metrics was taken from Lawton, G., August 22, 2023, "Eight areas for creating and refining generative AI metrics," TechTarget, www.techtarget.com/searchenterpriseai/feature/Areas-for-creating-and-refining-generative-AI-metrics?utm_campaign=20230824_ERU-ACTIVE_WITHIN_90_DAYS&utm_medium=email&utm_source=SGERU&source_ad_id=366549253&src=15003291&asrc=EM_SGERU_275384501.
3 The Fréchet Inception Distance (FID) is a metric used to evaluate the quality of images generated by machine-learning models, particularly generative adversarial networks (GANs). It compares the statistical similarity between features extracted from real images and those from generated images using a pre-trained Inception network. Lower FID scores indicate closer resemblance between real and generated images, with lower scores reflecting higher quality.
4 The F1 score is a metric used to evaluate the accuracy of a classification model by balancing precision (how many predicted positive cases are correct) and recall (how many actual positive cases are correctly identified). It is the harmonic mean

of precision and recall, ranging from 0 to 1, where a higher score indicates better performance. The F1 score is particularly useful when the data is imbalanced, meaning one class has significantly more examples than the other.

5 In the context of the BLEU metric, "n-gram levels" refer to sequences of n words used to evaluate the similarity between the machine-translated text and the reference text. For example, a one-gram considers single words, a two-gram considers pairs of consecutive words, and so on. Higher n-gram levels capture more contextual information, helping assess fluency and coherence in translations.

6 The main point, overarching theme, essence – that's gist in a nutshell. The gist of gist, if you will. The gist of a conversation, argument, story, or what-have-you is what we rely on when the actual words and details are only imperfectly recalled, inessential, or too voluminous to recount in their entirety. (n.a.), October 17, 2024, "Gist." Merriam-Webster, Incorporated, www.merriam-webster.com/dictionary/gist#:~:text=The%20main%20point%2C%20overarching%20theme,to%20recount%20in%20their%20entirety.

7 ROUGE-N is a type of ROUGE score that measures the overlap of n-grams (continuous sequences of n words) between a machine-generated text and a reference text. Commonly used for unigram (ROUGE-1) and bigram (ROUGE-2) comparisons, ROUGE-N assesses the recall of key words and phrases, helping evaluate the content similarity between the generated summary and the reference summary.

8 ROUGE-L is a type of ROUGE score that measures the longest common subsequence (LCS) between the machine-generated text and the reference text. It focuses on capturing sentence-level structure and sequence alignment, assessing how well the summary maintains the order of key phrases compared with the human reference. This makes ROUGE-L particularly useful for evaluating the coherence and fluency of summaries.

Chapter 4

Auditing AI KPIs

A key performance indicator (KPI) is a metric of progress toward intended results. Schrage and Kiron go a step further, defining KPIs as "the quantifiable measures an organization uses to determine how well it meets its declared operational and strategic goals.[1]

KPIs are generated by identifying and aligning measurable metrics with organizational goals, objectives, and desired outcomes. They are typically developed through collaboration among stakeholders, leveraging historical data, current performance metrics, and forward-looking strategies.

KPIs may be separated into four distinct categories based on their objective. See Table 4.1.

The following are five significant KPIs with both business and IT relevance that apply to AI projects that should be reviewed as part of an internal audit or risk assessment of an organization's AI systems, tools, applications, and usage.

Table 4.1 Types of KPIs

KPI Type	*Objective*	*Example*
Strategic	Focus on long-term goals and the overall vision of the organization. They measure progress toward achieving significant outcomes, such as market share growth, brand awareness, or customer satisfaction.	Market share percentage Customer lifetime value Employee engagement score
Tactical	Support strategic goals but are more short-term and specific. They are focused on departmental or project-based performance.	Sales conversion rate Average response time to customer inquiries Social media engagement rate
Operational	Measure the efficiency and effectiveness of daily operations, focusing on processes, quality, and productivity.	Order fulfillment time Machine uptime percentage
Financial	Focus on the fiscal health of an organization, including profitability, cash flow, and cost management.	Gross profit margin Operating cash flow Debt-to-equity ratio

DOI: 10.1201/9781003646501-5

1. Model Accuracy Rate
   Measures the percentage of correct predictions or classifications made by the AI system. A high accuracy rate ensures reliability and alignment with business objectives.

$$\text{Accuracy} = \frac{\text{Correct Predictions}}{\text{Total Predictions}} \times 100$$

   - Business Relevance: Directly impacts business outcomes, such as customer satisfaction, operational efficiency, or risk mitigation.
   - IT Relevance: Indicates the effectiveness and reliability of the underlying algorithms and data processing.
   - Example: For a fraud detection AI tool, the accuracy metric shows the proportion of fraudulent transactions correctly identified versus the total transactions analyzed.
2. Mean Time to Detection (MTTD) for Anomalies
   - Description: The average time it takes for the AI system to detect anomalies or potential issues.
   - Business Relevance: Minimizes the time window for financial losses, operational inefficiencies, or security breaches.
   - IT Relevance: Reflects the system's ability to process data efficiently and deliver actionable insights in real time.
   - Example: In cybersecurity AI, MTTD measures how quickly the system detects unusual login activity or data exfiltration attempts.
3. Bias and Fairness Metrics
   - Description: Evaluates the extent to which the AI model's predictions are equitable across different demographic or user groups.
   - Business Relevance: Ensures compliance with legal and ethical standards, avoiding reputational damage or regulatory penalties.
   - IT Relevance: Highlights potential issues in training data and model design that could lead to biased outcomes.
   - Example: For a recruitment AI system, fairness metrics might assess whether the system is equally accurate for applicants from different age groups or genders.
4. Resource Utilization Efficiency
   - Description: Measures the computational and storage resources consumed by AI models relative to their performance.
   - Business Relevance: Balances the cost of AI operations with the value they generate, ensuring financial sustainability.
   - IT Relevance: Reflects the efficiency of AI algorithms and infrastructure, highlighting opportunities for optimization.
   - Example: In an AI-driven customer support system, resource efficiency metrics could monitor cloud costs per query resolved.

5. Return on Investment (ROI) for AI Projects
   Measures the financial return generated by the AI system relative to its cost.

$$\text{ROI} = \frac{\text{Net Profit from AI}}{\text{Total Investment in AI}} \times 100$$

   - Business Relevance: Demonstrates the tangible value of AI projects to stakeholders and justifies further investments.
   - IT Relevance: Validates the strategic alignment of IT and AI initiatives with organizational goals.
   - Example: An AI-enabled inventory management system might report ROI by showing reduced stockouts and surplus costs.

The auditor, security, and risk professional should determine what if any, additional metrics are used to assist the organization in assessing if AI systems are meeting design, operational, and user objectives.

Additional metrics that are designed to assess AI system performance and, if not in use, may be recommended in the final assessment report include but are not limited to:

- Time to Deployment: Measures the time taken from developing an AI model to deploying it in a production environment. This reflects the agility and responsiveness of the AI team.
- Retraining Frequency: Tracks how often AI models require retraining to maintain performance. A low retraining frequency suggests robustness, while frequent retraining may indicate instability in the data or model.
- Inference Latency: Measures the time taken by the AI system to process input and generate output. Low latency is critical for real-time AI applications (e.g., fraud detection recommendation systems). Typically measured in milliseconds or seconds.
- Data Processing Throughput: Tracks the volume of data the AI system can process in a given time frame. Higher throughput indicates robust operational capacity. The unit of measure is typically data points or transactions per second.

AUDITING BUSINESS-RELATED AI KPIs

The following questions should be asked in the process of auditing, assessing, and evaluating AI systems.

1. Model Accuracy
   1. How does the organization define acceptable accuracy levels for its AI models, and how are these levels aligned with business objectives?

2. How does the organization ensure accuracy metrics are not influenced by biased or incomplete training data?
    3. How does the organization handle scenarios where model accuracy drops below predefined thresholds?
    4. What data sources are used to measure model accuracy, and how is the quality of this data ensured?
    5. What processes are in place to regularly evaluate and validate the accuracy of AI models post-deployment? How are these processes documented?
2. Mean Time to Detection (MTTD) for Anomalies
    1. How are detected anomalies prioritized, escalated, and resolved to minimize operational impact?
    2. How does the organization define and monitor acceptable thresholds for anomaly detection times in AI systems?
    3. How does the organization evaluate whether MTTD meets the needs of both IT and business stakeholders?
    4. What steps are taken to improve the speed and reliability of anomaly detection in critical AI applications?
    5. What tools and processes are in place to measure and analyze trends in MTTD over time?
3. Bias and Fairness Metrics
    1. How does the organization address fairness concerns raised by end-users or other stakeholders?
    2. How does the organization ensure that the data used to train AI models is representative and unbiased?
    3. How does the organization measure fairness across different demographic or stakeholder groups impacted by AI systems?
    4. What mechanisms exist to review and update fairness metrics as societal or regulatory expectations evolve?
    5. What processes are in place to identify and mitigate biases in AI model predictions or outcomes?
4. Resource Utilization Efficiency
    1. How does the organization ensure scalability in resource utilization as AI projects grow or change?
    2. How does the organization evaluate the cost-effectiveness of the AI infrastructure it uses (e.g., cloud vs on-premises)?
    3. How does the organization track and evaluate the computational and storage resources consumed by AI systems?
    4. What measures are taken to balance resource efficiency with sustainability goals, such as reducing energy consumption?
    5. What strategies are in place to optimize the efficiency of AI models while maintaining performance levels?
5. Return on Investment (ROI) for AI Projects
    1. How does the organization evaluate whether AI investments align with broader strategic goals?

2. How does the organization incorporate lessons learned from previous AI projects to improve ROI in future initiatives?
3. How does the organization measure the financial and non-financial benefits of AI projects relative to their costs?
4. What metrics are used to define the success of AI projects, and how are these metrics reviewed over time?
5. What processes are in place to identify and address underperforming AI initiatives that fail to deliver expected ROI?

NOTE

1 M. Schrage and D. Kiron, June 2018, "Leading With Next-Generation Key Performance Indicators," MIT Sloan Management Review, https://sloanreview.mit.edu/wp-content/uploads/2018/06/9dbcc7b64e.pdf.

Part 2

AI Governance and Oversight

Chapter 5

AI Governance and Oversight

WHAT IS ARTIFICIAL INTELLIGENCE GOVERNANCE?

Artificial intelligence (AI) governance is the legal framework for ensuring AI and machine-learning technologies are researched and developed to help humanity adopt and use these systems in ethical and responsible ways. AI governance aims to close the gap that exists between accountability and ethics in technological advancement.

The main focus of AI governance is on AI as it relates to justice, data quality, and autonomy. Overall, AI governance determines how much of daily life algorithms can shape and who monitors how AI functions.

Some key areas governance addresses include the following:

- AI safety.
- Sectors appropriate for AI automation.
- Legal and institutional structures around AI use and technology.
- Control and access to personal data.
- Moral and ethical questions related to AI.[1]

This component (AI governance) includes the structures, processes, and procedures needed to guide, manage, and oversee the organization's AI efforts aimed at achieving its goals. The right formality and structure of AI governance will differ depending on each company's unique circumstances and qualities.

AI governance addresses accountability and oversight, as well as whether personnel in charge of AI have the requisite skills and experience to monitor its use and whether its AI operations represent its values. Given the advances in AI's influence, it is vital that related activities and decisions are consistent with the organization's ethical, social, and legal obligations.[2]

DOI: 10.1201/9781003646501-7

WHY IS AI GOVERNANCE NECESSARY AND IMPORTANT?

AI governance is necessary and important because it ensures the responsible, ethical, and secure development, deployment, and use of AI systems.

The following key reasons highlight the importance of proactive AI governance.

- Ethical Considerations
  AI governance promotes fairness, accountability, and transparency, helping organizations address biases, discrimination, and ethical dilemmas that can arise from AI decision-making.
- Regulatory Compliance
  It helps organizations adhere to legal and regulatory requirements, such as data privacy laws (e.g., GDPR, CCPA), reducing the risk of legal consequences and fines.
- Risk Management
  Effective governance identifies, assesses, and mitigates risks related to AI systems, including security vulnerabilities, data breaches, and unintended consequences.
- Trust and Reputation
  By demonstrating a commitment to ethical AI practices, organizations can build trust with stakeholders, including customers, employees, and regulators, and maintain a positive reputation.
- Bias Mitigation
  Governance assists in ensuring that mechanisms are in place to identify and mitigate biases in AI systems, ensuring decisions are fair and equitable for all users.
- Accountability
  Clear governance structures establish accountability for AI outcomes, helping to address concerns about the lack of human oversight in automated decision-making.
- Alignment with Organizational Goals
  AI governance ensures that AI initiatives align with an organization's broader goals, values, and risk appetite, maximizing their strategic value.
- Sustainability
  Governance frameworks promote the responsible use of resources and consideration of AI's long-term impact on society, the environment, and the workforce.
- Innovation with Control
  It allows organizations to innovate with AI while maintaining control, avoiding risks that could derail projects or harm stakeholders.
- Preparation for Future Regulations
  Proactive governance positions organizations to adapt to emerging AI-related regulations and standards, ensuring continued compliance and operational resilience.

AUDITING AI GOVERNANCE AND OVERSIGHT

The following questions may be incorporated into a broader audit assessment of the organization's governance and oversight of the acquisition, deployment, and use of AI.

1. Can you describe the governance framework in place to ensure effective oversight of AI initiatives within your organization?
2. Can you provide examples of how governance practices have been applied to address challenges or risks associated with AI within your organization?
3. Has the organization defined clear goals and objectives for the AI system to ensure that intended outcomes are achieved?
4. How are conflicts of interest managed within the governance structure related to AI decision-making and oversight?
5. How does your organization define the roles and responsibilities for overseeing AI-related activities?
6. How does your organization ensure accountability for AI outcomes, particularly in the case of adverse or unintended consequences?
7. How does your organization ensure that AI-related activities align with broader corporate governance policies and strategies?
8. What is the process for escalating AI-related issues to senior management or the board of directors?
9. What mechanisms are used to ensure that those overseeing AI activities have the necessary expertise and resources?
10. What processes are in place to monitor and review the performance and compliance of AI systems regularly?
11. What steps are taken to ensure that the governance framework for AI adapts to emerging risks, regulatory changes, or technological advancements?

GAO ARTIFICIAL INTELLIGENCE ACCOUNTABILITY FRAMEWORK – AUDITING AI GOVERNANCE

The US Government Accountability Office (GAO), in developing the Artificial Intelligence Accountability Framework, created a set of specific audit questions aimed at AI governance at the organizational level.

The objective of these questions is to:

- Define clear goals and objectives for the AI system to ensure intended outcomes are achieved.
- Define clear roles, responsibilities, and delegation of authority for the AI system to ensure effective operations, timely corrections, and sustained oversight.

- Demonstrate a commitment to values and principles established by the organization to foster public trust in the responsible use of the AI system.
- Recruit, develop, and retain personnel with multidisciplinary skills and experiences in the design, development, deployment, assessment, and monitoring of AI systems.
- Include diverse perspectives from a community of stakeholders throughout the AI lifecycle to mitigate risk.
- Implement an AI-specific risk management plan to systematically identify, analyze, and mitigate risks.
- Establish and document technical specifications to ensure the AI system meets its intended purpose.
- Ensure the AI system complies with relevant laws, regulations, standards, and guidance.
- Promote transparency by enabling external stakeholders to access information on the design, operation, and limitations of the AI system.

Auditing AI Governance – GAO AI Accountability Framework Questions

The following is a list of audit questions designed to assess an organization's ability to meet the above AI governance objectives.

1. What goals and objectives does the organization expect to achieve by designing, developing, and/or deploying the AI system?
2. To what extent do stated goals and objectives represent a balanced set of priorities and adequately reflect stated values?
3. How does the AI system help the organization meet its goals and objectives?
4. To what extent does the organization communicate its AI strategic goals and objectives to the community of stakeholders?
5. To what extent does the organization have the necessary resources – funds, personnel, technologies, and time frames – to achieve the goals and objectives outlined for designing, developing, and deploying the AI system?
6. To what extent does the organization consistently measure progress toward stated goals and objectives?
7. What are the roles, responsibilities, and delegation of authorities of personnel involved in the design, development, deployment, assessment, and monitoring of the AI system?
8. To what extent has the organization clarified the roles and responsibilities, and delegated authorities to relevant stakeholders?
9. How does the organization demonstrate a commitment to stated values and principles?
10. To what extent has the organization operationalized its stated core values and principles for the AI system?

11. What policies has the organization developed to ensure the use of the AI system is consistent with its stated values and principles?
12. To what extent do these policies foster public trust and confidence in the use of the AI system?
13. How does the organization determine the necessary skills and experience needed to design, develop, deploy, assess, and monitor the AI system?
14. What efforts has the organization undertaken to recruit, develop, and retain competent personnel?
15. What efforts has the organization undertaken to recruit, develop, and retain a workforce with backgrounds, experience, and perspectives that reflect the community impacted by the AI system?
16. How does the organization assess whether personnel have the necessary skills, training, resources, and domain knowledge to fulfill their assigned responsibilities?
17. What factors were considered when identifying the community of stakeholders involved throughout the lifecycle?
18. Which stakeholders did the organization include throughout the design, development, deployment, assessment, and monitoring lifecycle?
19. To what extent do the teams responsible for developing and maintaining the AI system reflect diverse opinions, backgrounds, experiences, and perspectives?
20. What specific perspectives did stakeholders share, and how were they integrated across the design, development, deployment, assessment, and monitoring of the AI system?
21. To what extent has the organization addressed stakeholder perspectives on the potential negative impacts of the AI system on end-users and impacted populations?
22. To what extent has the organization developed an AI-specific risk management plan to systematically identify, analyze, and mitigate known and unknown operational, technical, as well as societal risks associated with the AI system?
23. To what extent has the organization defined its risk tolerance for using the AI system?
24. To what extent does the plan specifically address risks associated with acquisition, procurement of packaged software from vendors, cybersecurity controls, computational infrastructure, data, data science, deployment mechanics, and system failure?
25. What challenge/constraint is the AI system intended to solve?
26. To what extent has the organization clearly defined technical specifications and requirements for the AI system?
27. How do the technical specifications and requirements align with the AI system's goals and objectives?
28. What justifications, if any, has the organization provided for the assumptions, boundaries, and limitations of the AI system?

29. To what extent has the organization identified the relevant laws, regulations, standards, and guidance, applicable to the AI system's use?
30. How does the organization ensure that the AI system complies with relevant laws, regulations, standards, federal guidance, and policies?
31. To what extent is the AI system in compliance with applicable laws, regulations, standards, federal guidance, and organization policies?
32. What type of information is accessible on the design, operations, and limitations of the AI system to external stakeholders, including end-users, consumers, regulators, and individuals impacted by the use of the AI system?
33. How easily accessible and current is the information available to external stakeholders?
34. To what extent is this information sufficient and appropriate to promote transparency?[3]

NOTES

1 Lewis, S., Barney, N., and Hashemi-Pour, C., August 11, 2023, "Artificial Intelligence (AI) Governance," TechTarget, www.techtarget.com/searchenterpriseai/definition/AI-governance.
2 Zygoulis, F., August 14, 2024, "Artificial Intelligence and the Future of the Internal Audit Function: A Systematic Literature Review," https://ssrn.com/abstract=4927352.
3 The source for these questions resides with the GAO analysis of US and foreign government, industry, academic, and nonprofit organization documents, interviews, and forum discussions. Ariga, T., Persons, T., and Sanford, S., June 2021, "Artificial Intelligence Accountability Framework, An Accountability Framework for Federal Agencies and Other Entities," US Government Accountability Office (GAO), GAO-21–519SP, www.gao.gov/assets/gao-21–519sp.pdf.

Chapter 6

Auditing AI Accountability

WHAT IS ACCOUNTABILITY IN AI?

Accountability in artificial intelligence (AI) refers to the responsibility of AI developers, operators, and organizations to ensure that AI systems operate transparently, ethically, and fairly. This includes addressing potential biases, errors, and ethical issues that arise during the development and deployment of AI technologies.

1. How does the organization ensure governance and oversight of AI systems to maintain accountability?
2. What mechanisms are in place to ensure transparency in the decision-making processes of AI systems?
3. How does the organization identify, monitor, and mitigate biases in its AI systems?
4. What ethical guidelines does the organization follow in the development and deployment of AI tools?
5. How does the organization track, report, and rectify errors made by AI systems?
6. How are stakeholders (e.g., users, customers, regulators) involved in the design and review of AI systems to ensure accountability?
7. How is the development process and decision logic of AI systems documented and communicated to relevant parties?
8. What risk management strategies are in place to address potential legal and reputational risks associated with AI?
9. How are AI systems validated and tested to ensure they function correctly and ethically?
10. What processes are in place for the continuous monitoring and improvement of AI systems to ensure ongoing accountability and alignment with ethical standards?

DOI: 10.1201/9781003646501-8

Accountability – Decision Trees

Decision Trees are a type of machine learning algorithm used for classification and regression tasks. They model decisions and their possible consequences in a tree-like structure. Decision trees are particularly useful in AI applications for ensuring accountability and traceability because they provide a clear and interpretable path of decision-making, allowing auditors to trace back decisions to specific attributes and values.

1. How does the AI system document and present the decision paths taken within decision trees?
2. What criteria are used for selecting attributes at each node in the decision tree, and how is this process documented?
3. How does the decision tree algorithm address and mitigate potential biases in the attribute selection process?
4. What measures are in place to ensure the integrity and quality of the data used to train the decision trees?
5. How does the organization ensure that the decision trees remain interpretable and understandable to non-technical stakeholders?
6. What processes are followed for identifying, analyzing, and correcting errors in the decisions made by the AI system?
7. How does the organization track changes and updates to the decision tree models over time?
8. How are the decision-making processes and criteria of the decision trees documented and communicated to relevant parties?
9. How does the organization assess the impact of decisions made by the AI system on different stakeholder groups?
10. How does the organization ensure that the decision trees comply with relevant ethical guidelines and regulatory requirements?

Accountability – Layer-Wise Relevance Propagation

Layer-Wise Relevance Propagation (LRP) is an interpretability method used in AI, particularly in neural networks, to explain the decisions made by the model. It works by propagating the prediction score of a neural network backward through the layers to distribute the score across the input features, indicating their relevance to the final decision. This method helps in understanding which features contributed most to a given prediction, thereby enhancing the transparency and traceability of AI systems.

1. How does the organization implement LRP to calculate the relevance of each input feature for the decisions made by the AI model?
2. How are the LRP results presented to ensure they are interpretable and understandable by non-technical stakeholders?
3. What steps are taken to ensure that the application of LRP provides a transparent view of the model's decision-making process?

4. How does LRP help in detecting and mitigating biases in the AI model's decisions?
5. How are the LRP methods and results documented and maintained for accountability purposes?
6. What processes are in place to validate the accuracy and reliability of the relevance scores generated by LRP?
7. How does the application of LRP impact the overall performance and accuracy of the AI model?
8. How are users and stakeholders involved in the interpretation and evaluation of the LRP results?
9. How does the use of LRP align with the organization's ethical guidelines for AI development and deployment?
10. What measures are taken to ensure that the application of LRP complies with relevant regulatory requirements and industry standards for AI accountability and traceability?

Accountability – Saliency Maps

Saliency Maps are visualization tools used in AI, especially in neural networks, to highlight the regions of input (such as an image) that are most important for the model's prediction. By showing which parts of the input contribute most to the decision, saliency maps help in interpreting the model's behavior, thereby enhancing accountability and traceability in AI applications.

1. How are saliency maps generated and presented to ensure that they are easily interpretable by non-technical stakeholders?
2. How does the organization ensure that the saliency maps accurately highlight the features most relevant to the model's decisions?
3. In what ways do saliency maps contribute to the transparency of the AI model's decision-making process?
4. How are saliency maps used to detect and address potential biases in the AI model's predictions?
5. How are the methods and results of generating saliency maps documented for future reference and accountability?
6. What processes are in place to validate the accuracy and reliability of the insights provided by saliency maps?
7. How is feedback from users and stakeholders incorporated into the evaluation of saliency maps?
8. How do saliency maps impact the understanding of model performance and error analysis?
9. How do saliency maps help in ensuring that the AI system's decisions align with ethical guidelines?
10. What measures are taken to ensure that the use of saliency maps complies with relevant regulatory requirements and industry standards for AI accountability and traceability?

Accountability – LIME, SHAP, and LINGO

LIME (Local Interpretable Model-agnostic Explanations) provides human-understandable explanations for predictions made by any black-box AI model, especially complex models like neural networks, ensemble models, etc. LIME explains an individual prediction (local explanation), rather than trying to explain the whole model globally. It works model-agnostically, meaning it can be used on any classifier or regressor (you don't need to know the model internals).

SHAP (SHapley Additive exPlanations) tells you for this specific prediction, which features mattered most, and how much they pushed the result up or down. Based on game theory (Shapley values), SHAP assigns a fair value to each feature, representing its contribution to the model's output. SHAP is useful as it:

- Makes AI model predictions interpretable and transparent.
- Helps build trust with users and regulators.
- Aids in debugging and identifying biases in models.

LINGO (Local Interpretable Model-agnostic Explanations with Global Optimization) is an algorithm designed to enhance the interpretability and explicability of AI models. LINGO works by generating local explanations for individual predictions and optimizing these explanations globally to ensure consistency and coherence across the model's predictions. This method helps users understand how specific inputs lead to particular outputs, thereby making AI models more transparent and accountable. This is part of the growing field of XAI (Explainable AI).

Auditing General AI Accountability

1. How are the ethical considerations of AI models, including fairness, transparency, and bias mitigation, documented and reviewed across the model lifecycle?
2. How does the organization ensure that AI models remain accountable when deployed in environments or contexts different from those in which they were trained?
3. How is accountability for AI models defined and operationalized within the organization's AI governance framework?
4. What mechanisms are in place to document model development decisions, training data provenance, and model versioning?
5. What processes are used to monitor and update models post-deployment to maintain accountability and prevent model drift?

Auditing Accountability Tools (LINGO, LIME, SHAP)

1. How are LIME-generated explanations validated for accuracy and reliability, and how is this validation process documented?
2. How are LINGO, LIME, and SHAP explanations incorporated into end-user documentation or interfaces to enhance transparency?
3. How does the organization assess and manage the risk of explanations exposing sensitive information or enabling reverse engineering of proprietary models?
4. How does the organization assess the effectiveness of LINGO's explanations in helping stakeholders (technical and non-technical) understand and trust the AI model's predictions?
5. How does the organization document and communicate the insights and limitations derived from LINGO explanations to regulators, internal auditors, and end users?
6. Through what methods does the organization evaluate the risk of explanations (produced by LINGO, LIME, SHAP) providing false confidence in model decisions that may be inaccurate or biased?
7. How does the organization leverage SHAP outputs to identify potential biases in model predictions, and what actions are taken to address any issues discovered?
8. How does the use of LINGO, LIME, and SHAP align with industry best practices and applicable regulatory requirements for AI transparency and accountability?
9. How is LINGO applied to enhance model interpretability, and what processes are followed to ensure that local and global explanations remain accurate and consistent?
10. How is SHAP used to quantify feature contributions in model outputs, and what criteria are applied to determine the trustworthiness of these explanations?
11. In what ways has the application of LIME supported accountability by making individual model predictions more transparent to relevant stakeholders?
12. What controls are implemented to detect and mitigate the risk of explanation tools (like LINGO, LIME, SHAP) failing to keep pace with changes in the underlying model (e.g., after retraining or updates)?
13. What governance processes are in place to ensure that explanations generated by LINGO, LIME, or SHAP remain current and reflective of the model as it evolves?
14. What procedures are in place to ensure that explanation tools themselves are validated, tested, and audited regularly to maintain their reliability and minimize operational risks?

15. What processes are in place to identify and assess risks related to model misinterpretation or misuse of LINGO, LIME, or SHAP explanations by stakeholders?

Additional tools may be used to assess an AI model's behavior, with the specific objective of enhancing the overall transparency, interpretability, explicability, auditability, and accountability of AI systems.

See Appendix 5 for a list and summary of some of these additional tools.

Chapter 7

Auditing the Trustworthiness of AI

> Designing systems that are transparent about the use of AI and are accountable for their outcomes is critical. AI systems must function properly and in a secure and safe manner.[1]

WHAT IS TRUSTWORTHY AI?

Trustworthy AI refers to artificial intelligence systems designed, developed, and deployed to ensure they are lawful, ethical, and robust. These systems must respect fundamental rights, operate transparently, ensure fairness, and prioritize safety and reliability while being aligned with societal and environmental well-being. Trustworthy AI incorporates principles such as accountability, human oversight, data privacy, and the minimization of bias, ensuring the technology is beneficial and safe for all stakeholders.

Characteristics of trustworthy AI systems include valid and reliable, safe, secure and resilient, accountable and transparent, explainable and interpretable, privacy-enhanced, and fair with harmful bias managed. Valid and reliable is a necessary condition of trustworthiness and is shown as the basis for other trustworthiness characteristics. Accountable and transparent is shown as a vertical box because it relates to all other characteristics; see Figure 7.1.

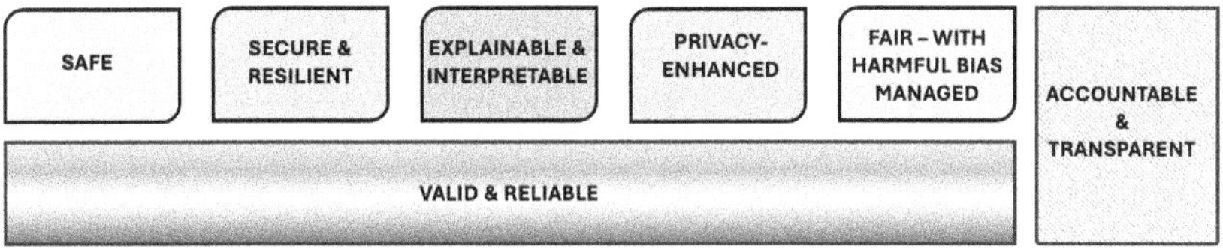

Figure 7.1 Characteristics of Trustworthy AI Systems.[2]

DOI: 10.1201/9781003646501-9

Creating trustworthy AI requires balancing each of these characteristics based on the AI system's context of use. While all characteristics are socio-technical system attributes, accountability and transparency also relate to the processes and activities internal to an AI system and its external setting. Neglecting these characteristics can increase the probability and magnitude of negative consequences.[3]

AUDITING THE TRUSTWORTHINESS OF AI – OBJECTIVES

Auditing the trustworthiness of AI involves systematically evaluating AI systems to ensure they align with ethical, legal, and technical standards of trustworthiness. This process includes assessing key aspects such as fairness, transparency, accountability, data privacy, robustness, and the system's alignment with societal values and laws. Auditing seeks to identify potential biases, unintended consequences, security vulnerabilities, and compliance gaps throughout the AI lifecycle.

The goal is to build confidence among stakeholders by demonstrating that the AI system operates reliably, respects human rights, and minimizes risks to individuals and societies. The process typically involves reviewing system design, data usage, decision-making processes, and operational impacts, along with ensuring mechanisms for redress and accountability are in place. By addressing these elements, AI audits foster trust and ethical integration of AI technologies into various applications.

These seven critical areas ensure that the AI system's entire lifecycle meets the key requirements for trustworthy AI:

1. Human agency and oversight.
2. Technical robustness and safety.
3. Privacy and data governance.
4. Transparency.
5. Diversity, non-discrimination, and fairness.
6. Environmental and societal well-being.
7. Accountability.

In addition, auditors should be mindful that the AI system, and the system's developers under review should also:

- Communicate clearly and proactively information to stakeholders about the AI system's capabilities and limitations, enabling realistic expectation setting and about how the requirements are implemented.
- Be transparent to end-users about the fact that they are dealing with an AI system.

- Facilitate the traceability and auditability of AI systems, particularly in critical contexts and situations.
- Involve stakeholders throughout the AI system's lifecycle. Foster training and education so that all stakeholders are aware of and trained in trustworthy AI.[4]

Critical Operational Elements of Trustworthiness

In his article "Auditing the Trustworthiness of an AI Algorithm," the author Juan identifies five critical operational features or elements that require audit attention and examination when performing an audit of the trustworthiness of an AI algorithm.

These five features are:

1. Model Validation and Performance Assessment
    - Proper internal validation frameworks (e.g., cross-validation) must be implemented to avoid data contamination, bias, and overfitting.
    - External validation through real-world field tests ensures the AI system performs accurately and robustly in operational conditions, including feedback from users.
2. Transparency and Explainability
    - Transparency in model operations and decisions is crucial for legal and ethical compliance.
    - Explainability methods must address both global (overall system behavior) and local (instance-specific decisions) levels, with techniques like SHAP or permutation importance used to clarify predictions.
3. Data Integrity and Privacy
    - High-quality data pipelines and adherence to privacy regulations (e.g., GDPR) are essential.
    - Audits should evaluate data-sampling methods, data minimization, proportionality, and transparency, especially when personal data is involved.
4. Fairness and Bias Mitigation
    - Fairness must be embedded into the model, especially in handling imbalanced datasets.
    - Auditors should ensure the system's fairness criteria are consciously chosen, balancing metrics like recall, precision, or prevalence, depending on the context.
5. Human Oversight and Ethical Robustness
    - AI systems should respect human autonomy, ensuring mechanisms for human oversight are in place.
    - Predictability, stability, and adherence to ethical principles like fairness, privacy, and accountability should be central throughout the system's lifecycle.[5]

These elements form a foundation for ensuring the AI system is lawful, ethical, and robust, aligning with the EU Ethics Guidelines for Trustworthy AI.

As the public gains knowledge and understanding of issues around the use and application of AI (including bias, fairness, accountability, and responsibility) coupled with an increased awareness of data privacy, both public services and the private sector will have to become more accountable if they win public trust and secure the vital public "license to operate." Reputational damage as a result of insufficient or ineffective data and AI governance can cause significant harm to a business.[6]

THE EU GUIDELINES FOR TRUSTWORTHY AI

The EU Guidelines for Trustworthy AI, developed by the European Commission's High-Level Expert Group on AI, provide a framework for ensuring that AI systems are ethical, lawful, and robust. These guidelines identify seven key requirements for AI systems: human agency and oversight, technical robustness and safety, privacy and data governance, transparency, diversity and non-discrimination, societal and environmental well-being, and accountability. By meeting these requirements, AI systems can support human rights, prevent harm, and foster trustworthiness across their lifecycle.

The guidelines aim to operationalize ethical principles like respect for autonomy, prevention of harm, fairness, and explicability, ensuring that AI technologies enhance societal well-being while minimizing risks. The EU Guidelines encourage a holistic approach to AI governance, balancing technical and social factors to promote responsible AI innovation that aligns with European values and protects fundamental rights.

Auditing AI Trustworthiness: The EU Guidelines for Trustworthy AI

Incorporating an audit review of the EU Guidelines for Trustworthy AI is crucial to assess whether an organization's AI initiatives align with established ethical, legal, and robustness standards. This review ensures that AI systems are technically sound, ethically responsible, and legally compliant. Auditing against these guidelines helps identify gaps or risks in transparency, accountability, and data governance, allowing organizations to proactively address potential issues before they lead to harm or regulatory non-compliance.

Furthermore, including these guidelines in an AI audit supports organizational trust and credibility, as it demonstrates a commitment to ethical and responsible AI use. Organizations can enhance their reputation, build public confidence, and potentially mitigate reputational and operational risks associated with AI technologies by moving toward compliance with and embracing the EU's principles for trustworthy AI. This comprehensive approach to AI governance is essential for long-term, sustainable AI deployment that aligns with both organizational goals and societal expectations.

THE ETHICS GUIDELINES FOR TRUSTWORTHY AI

The Ethics Guidelines for Trustworthy AI, prepared by the High-Level Expert Group on Artificial Intelligence (AI HLEG) of the European Commission, outline a framework to ensure AI systems are ethical, lawful, and robust. The guidelines emphasize that trustworthy AI is built on three foundational components: it must comply with applicable laws, adhere to ethical principles, and be technically robust to prevent unintended harm.

These principles are supported by key requirements, such as human agency and oversight, technical robustness and safety, privacy and data governance, transparency, diversity and fairness, societal well-being, and accountability.

The document serves as a practical guide for AI developers, deployers, auditors, and users, offering a checklist to operationalize these principles across the lifecycle of AI systems. The guidelines aim to align AI innovation with European values, fostering trust while mitigating risks.

By addressing challenges like bias, data misuse, and societal impact, the framework seeks to ensure that AI enhances human rights, democracy, and social equity.

AUDITING TRUSTWORTHY AI – AI HLEG FRAMEWORK QUESTIONS

Transparency

1. How does the organization communicate the presence of AI systems to users and ensure they are aware of interacting with AI?
2. How does the organization ensure that the decision-making processes of AI systems are transparent and traceable?
3. How is the rationale behind AI-generated decisions explained to end-users, particularly in high-stakes contexts?
4. What measures are in place to document and track the data and algorithms used in AI systems to ensure auditability?
5. What steps are taken to communicate the capabilities, limitations, and purpose of AI systems to users and stakeholders?

Traceability

1. What measures has the organization put in place to ensure traceability, including the documentation of methods used for designing and developing the algorithmic system?

2. In the case of a rule-based AI system, how has the organization documented the method of programming or the process by which the model was built?
3. For a learning-based AI system, how has the organization documented the method of training the algorithm, including the selection and gathering of input data, and the steps taken during this process?
4. What methods has the organization used to test and validate the algorithmic system?
5. In the case of a rule-based AI system, what scenarios or cases has the organization documented to test and validate the system?
6. For a learning-based model, what information about the data used for testing and validation has the organization documented?
7. What outcomes of the algorithmic system has the organization documented?
8. How has the organization documented the decisions made by the algorithm, as well as potential alternative decisions that could result from different cases, such as for other subgroups of users?

Explainability

1. How has the organization assessed the extent to which the decisions and outcomes made by the AI system can be understood?
2. How does the organization ensure that explanations for why a system makes certain choices and outcomes are understandable to all users who may require an explanation?
3. To what degree has the organization evaluated how the system's decisions influence its decision-making processes?
4. Why has the organization chosen to deploy this particular system in this specific area, and how has this been assessed?
5. How has the organization evaluated the business model concerning this system, including how it creates value for the organization?
6. How has the organization designed the AI system with interpretability in mind from the outset?
7. What efforts has the organization made to research and use the simplest and most interpretable model possible for the specific application?
8. How does the organization assess its ability to analyze training and testing data, and how does it ensure this data can be updated or changed over time?
9. What steps has the organization taken to examine the interpretability of the model after training and development, and does it have access to the model's internal workflow?

Communication

1. How does the organization communicate to end-users through disclaimers or other means that they are interacting with an AI system and not a human? How has the organization labeled the AI system as such?

2. What mechanisms has the organization put in place to inform users about the reasons and criteria behind the AI system's outcomes?
3. How does the organization ensure that this information is clearly and intelligibly communicated to intended users?
4. What processes has the organization established to consider user feedback and adapt the system accordingly?
5. How has the organization communicated potential or perceived risks of the AI system, such as bias?
6. Depending on the use case, how has the organization ensured communication and transparency toward other audiences, such as third parties or the general public?
7. How has the organization clarified the purpose of the AI system and who or what may benefit from its product or service?
8. How has the organization specified and communicated the usage scenarios for the product, including alternative forms of communication to ensure they are understandable and appropriate for target users?
9. How has the organization addressed human psychology and potential limitations, such as risks of confusion, confirmation bias, or cognitive fatigue, depending on use case?
10. In what ways has the organization communicated the characteristics, limitations, and potential shortcomings of the AI system, both during the development and deployment phases, to relevant stakeholders or end-users?

Technical Robustness and Safety

1. How are fallback mechanisms or fail-safes integrated into AI systems to mitigate risks in case of malfunctions?
2. How does the organization ensure that AI systems remain secure and resilient to adversarial attacks or manipulation?
3. How does the organization monitor and test the accuracy and reliability of AI outputs over time?
4. What is the process for identifying and addressing any potential failures or unintended consequences of AI systems?
5. What measures are implemented to ensure the technical robustness and resilience of AI systems against potential attacks?

Resilience to Attack and Security

1. What potential forms of attack has the organization assessed for vulnerabilities in the AI system?
2. How has the organization considered different types and natures of vulnerabilities, such as data pollution, threats to physical infrastructure, and cyber-attacks?
3. What measures or systems has the organization put in place to ensure the integrity and resilience of the AI system against potential attacks?

4. How has the organization assessed the AI system's behavior in unexpected situations and environments?
5. To what extent has the organization considered whether the system could be dual-use, and what preventative measures have been taken to address this risk, such as not publishing the research or refraining from deploying the system?

Fallback Plan and General Safety

1. How has the organization ensured that the AI system has a sufficient fallback plan in case of adversarial attacks or other unexpected situations (e.g., technical switching procedures or involving a human operator before proceeding)?
2. How has the organization assessed the level of risk raised by the AI system in this specific use case?
3. What processes has the organization implemented to measure and assess risks and safety?
4. What information has the organization provided in case of risks to human physical integrity?
5. Has the organization considered an insurance policy to address potential damages caused by the AI system? If so, what does it entail?
6. How has the organization identified and addressed potential safety risks from foreseeable uses of the technology, including accidental or malicious misuse? Is there a mitigation or management plan for these risks?
7. How has the organization assessed whether the AI system might cause harm to users or third parties? What factors, such as likelihood, potential damage, impacted audience, and severity, were considered?
8. In cases where there is a risk of damage caused by the AI system, how has the organization addressed liability and consumer protection rules?
9. What potential impacts or safety risks to the environment or animals has the organization considered concerning the AI system?
10. Does the organization's risk analysis consider cybersecurity hazards or network vulnerabilities that could lead to unintentional behaviors or damage caused by the AI system?
11. How has the organization estimated the likely impact of system failures, including incorrect results, system unavailability, or societal harms such as discriminatory practices?
12. What thresholds and governance structures has the organization established for triggering alternative or fallback plans for the scenarios described?
13. How has the organization defined and tested fallback plans for the AI system to ensure readiness for failure scenarios?

Accuracy

1. What level and definition of accuracy has the organization assessed as necessary for the AI system in the context of its use case?
2. How does the organization measure and assure the accuracy of the AI system?
3. What measures has the organization implemented to ensure that the data used is comprehensive and up to date?
4. How has the organization assessed the need for additional data to improve accuracy or eliminate bias?
5. How has the organization evaluated the potential harm caused by inaccurate predictions made by the AI system?
6. What processes has the organization put in place to measure whether the system is making an unacceptable number of inaccurate predictions?
7. If inaccurate predictions occur, what steps has the organization established to resolve the issue effectively?

Reliability and Reproducibility

1. What strategies has the organization implemented to monitor and test that the AI system meets its goals, purposes, and intended applications?
2. How has the organization tested whether specific contexts or particular conditions need to be considered to ensure the reproducibility of the AI system?
3. What processes or methods has the organization established for verifying and measuring different aspects of reliability and reproducibility in the AI system?
4. How has the organization defined processes for identifying and describing when the AI system fails in certain settings?
5. In what ways has the organization documented and operationalized the processes for testing and verifying the reliability of AI systems?
6. What mechanisms or communication approaches have the organization implemented to assure end-users of the reliability of the AI system?

Privacy and Data Governance

1. How does the organization address data biases and inaccuracies that may affect AI system outcomes?
2. How does the organization protect the privacy of individuals in the design and deployment of AI systems?
3. How is access to data regulated, and who is authorized to handle sensitive or personal information within AI systems?
4. What data governance policies are in place to ensure the quality, integrity, and relevance of data used by AI systems?

5. What mechanisms are in place to prevent data misuse and ensure data is only used for its intended purpose?

Respect for Privacy and Data Protection

1. How has the organization established mechanisms that allow others to flag issues related to privacy or data protection in the AI system's processes of data collection (for training and operation) and data processing?
2. How has the organization assessed the type and scope of data in its data sets, such as whether they contain personal data?
3. What approaches has the organization considered to develop the AI system or train the model with minimal or no use of potentially sensitive or personal data?
4. What mechanisms has the organization built for providing notice and control over personal data depending on the use case, such as obtaining valid consent and allowing the possibility to revoke it when applicable?
5. What measures has the organization implemented to enhance privacy, such as encryption, anonymization, and aggregation?
6. Where a Data Privacy Officer exists, how has the organization involved this person at an early stage in the process?

Quality and Integrity of Data

1. How has the organization aligned its system with potentially relevant standards (e.g., ISO, IEEE) or widely adopted protocols for daily data management and governance?
2. What oversight mechanisms has the organization established for data collection, storage, processing, and use?
3. How has the organization assessed its control over the quality of external data sources used in the system?
4. What processes has the organization put in place to ensure the quality and integrity of its data? What additional processes have been considered, and how does the organization verify that its data sets have not been compromised or hacked?

Access to Data

1. What protocols, processes, and procedures have the organization followed to manage and ensure proper data governance?
2. How has the organization assessed who can access users' data and under what circumstances?
3. How does the organization ensure that individuals accessing the data are qualified, required to access it, and possess the necessary competencies to understand the details of the data protection policy?
4. What oversight mechanisms has the organization established to log when, where, how, by whom, and for what purpose data is accessed?

Diversity, Non-Discrimination, and Fairness

1. How are diverse perspectives and backgrounds considered in the design and development of AI systems?
2. How does the organization identify and mitigate biases in AI systems to ensure fair treatment across diverse user groups?
3. How does the organization involve stakeholders in assessing the fairness and inclusivity of AI systems?
4. What procedures are in place to ensure fairness in AI decision-making, particularly for marginalized or vulnerable populations?
5. What steps are taken to ensure that AI systems are accessible to individuals with varying abilities and characteristics?

Unfair Bias Avoidance

1. What strategies or procedures has the organization implemented to avoid creating or reinforcing unfair bias in the AI system, both in terms of input data and algorithm design?
2. How has the organization assessed and acknowledged possible limitations stemming from the composition of the datasets used?
3. How has the organization considered the diversity and representativeness of users in the data? Has it tested for specific populations or problematic use cases?
4. What technical tools has the organization researched and used to improve its understanding of the data, model, and performance?
5. What processes has the organization established to test and monitor for potential biases during the development, deployment, and use phases of the system?
6. Depending on the use case, how has the organization ensured a mechanism for flagging issues related to bias, discrimination, or poor performance of the AI system?
7. What clear steps and communication methods has the organization considered for how and to whom such issues can be raised?
8. Has the organization considered not only end-users, but also other parties potentially indirectly affected by the AI system?
9. How has the organization assessed whether decision variability could occur under the same conditions, and what possible causes have been identified?
10. In cases of variability, what measurement or assessment mechanisms have the organization established to evaluate the impact of such variability on fundamental rights?
11. What working definition of "fairness" has the organization applied in designing AI systems, and how was this definition chosen?
12. Is the organization's chosen definition of fairness commonly used? Have alternative definitions been considered before finalizing this one?
13. What quantitative analyses or metrics has the organization used to measure and test the applied definition of fairness?

14. What mechanisms has the organization established to ensure fairness in its AI systems, and have alternative mechanisms been considered?

Accessibility and Universal Design

1. How has the organization ensured that the AI system accommodates a wide range of individual preferences and abilities?
2. How has the organization assessed whether the AI system is usable by individuals with special needs or disabilities or those at risk of exclusion? How were these considerations designed into the system, and how are they verified?
3. What steps has the organization taken to ensure that information about the AI system is accessible to users of assistive technologies?
4. How has the organization involved or consulted communities with special needs during the development phase of the AI system?
5. How has the organization accounted for the impact of the AI system on the potential user audience?
6. Is the team involved in building the AI system representative of the target user audience, and does it also represent the wider population, including groups who might be tangentially impacted?
7. How has the organization assessed whether some persons or groups might be disproportionately affected by the negative implications of the AI system?
8. What feedback has the organization sought from other teams or groups that represent diverse backgrounds and experiences?

Stakeholder Participation

1. How has the organization considered mechanisms to include the participation of different stakeholders in the AI system's development and use?
2. What steps has the organization taken to pave the way for the introduction of the AI system by informing and involving impacted workers and their representatives in advance?

Human Agency and Oversight

1. How are users informed about the presence of AI systems and empowered to make informed choices about their interactions?
2. How does the organization assess the impact of AI on human rights, and what steps are taken to mitigate any identified risks?
3. How does the organization ensure AI systems support human autonomy and do not unduly influence or manipulate user behavior?
4. What mechanisms are in place to enable human oversight of AI systems, particularly for critical decision-making processes?
5. What procedures ensure that human operators can intervene or override AI systems when necessary?

Fundamental *Rights*

1. How has the organization carried out a fundamental rights impact assessment in use cases where there could be a negative impact on fundamental rights? What potential trade-offs between different principles and rights were identified and documented?
2. In what ways does the AI system interact with decision-making by human end-users (e.g., through recommended actions, decisions, or presenting options)?
3. How has the organization assessed the risk of the AI system potentially affecting human autonomy by unintentionally interfering with the end-user's decision-making process?
4. How has the organization considered whether the AI system should communicate to users that a decision, content, advice, or outcome is the result of an algorithmic decision?
5. In cases where the AI system features a chatbot or conversational system, how has the organization ensured that human end-users are made aware they are interacting with a non-human agent?

Human Agency

1. How has the organization considered task allocation between the AI system and human workers to ensure meaningful interactions and appropriate human oversight and control in work and labor processes?
2. In what ways does the AI system enhance or augment human capabilities?
3. What safeguards has the organization implemented to prevent overconfidence or overreliance on the AI system in work processes?

Human Oversight

1. How has the organization determined the appropriate level of human control for the specific AI system and use case?
2. What is the organization's approach to human control or involvement, if applicable? Who is designated as the "human in control," and what are the moments or tools for human intervention?
3. What mechanisms and measures has the organization implemented to ensure potential human control or oversight and to guarantee that decisions are taken under the overall responsibility of human beings?
4. What measures has the organization taken to enable audits and resolve issues related to governing AI autonomy?
5. In the case of a self-learning or autonomous AI system or use case, what specific mechanisms of control and oversight has the organization implemented?
6. What detection and response mechanisms have the organization established to identify and address potential issues or failures in the AI system?

7. How has the organization ensured the presence of a "stop button" or procedure to safely abort an operation when needed? Does this procedure halt the process entirely, in part, or delegate control to a human?

Societal and Environmental Well-Being

1. How are the ethical implications of AI system deployment on societal and democratic processes considered?
2. How does the organization assess the societal impact of AI systems on communities and societal structures?
3. How does the organization ensure that AI systems contribute positively to social well-being and avoid exacerbating inequalities?
4. What actions does the organization take to support AI initiatives that address global challenges, such as sustainability?
5. What measures are implemented to minimize the environmental footprint of AI development, deployment, and operations?

Sustainable and Environmentally Friendly AI

1. What mechanisms has the organization implemented to measure the environmental impact of the AI system's development, deployment, and use (e.g., energy consumption by data centers, type of energy used)?
2. What measures has the organization taken to reduce the environmental impact of the AI system's lifecycle?

Social Impact

1. In cases where the AI system directly interacts with humans, how has the organization assessed whether the AI system encourages humans to develop attachment and empathy toward it?
2. How has the organization ensured that the AI system signals that its social interaction is simulated and that it does not possess capacities for "understanding" or "feeling?"
3. How has the organization ensured that the social impacts of the AI system are well understood? For example, has the organization assessed potential risks of job loss or de-skilling of the workforce, and what steps have been taken to counteract such risks?

Society and Democracy

1. How has the organization assessed the broader societal impact of the AI system's use beyond individual end-users, including potentially indirectly affected stakeholders?

Accountability

1. How does the organization ensure that responsibility for AI outcomes is clearly assigned and understood by relevant stakeholders?

2. How does the organization handle and document any negative impacts or unintended consequences of AI system use?
3. What is the process for auditing AI systems, and how often are these audits conducted?
4. What mechanisms are in place to ensure accountability for AI decisions, particularly those with significant impacts on individuals or society?
5. What procedures are in place for redress when an AI system's decision causes harm or adversely impacts users?

Auditability

1. What mechanisms has the organization implemented to facilitate the AI system's auditability by internal and/or independent actors, such as ensuring traceability and logging of the system's processes and outcomes?

Minimizing and Reporting Negative Impact

1. How has the organization carried out a risk or impact assessment of the AI system that considers different stakeholders who are directly and indirectly affected?
2. What training and education frameworks has the organization implemented to develop accountability practices?
3. Which workers or branches of the team are involved in these initiatives, and do they extend beyond the development phase of the AI system?
4. Do the organization's training programs include instruction on the potential legal frameworks applicable to the AI system?
5. Has the organization considered establishing an "AI ethics review board" or a similar mechanism to discuss overall accountability and ethical practices, including potentially unclear or gray areas?
6. In addition to internal initiatives or frameworks for overseeing ethics and accountability, what forms of external guidance or auditing processes has the organization established?
7. What processes has the organization implemented for third parties (e.g., suppliers, consumers, distributors, vendors) or workers to report potential vulnerabilities, risks, or biases in the AI system or application?

Documenting Trade-Offs

1. What mechanisms has the organization established to identify relevant interests and values implicated by the AI system and potential trade-offs between them?
2. What processes does the organization use to decide on such trade-offs, and how does it ensure that trade-off decisions are properly documented?

Ability to Redress

1. What mechanisms has the organization established to allow for redress in cases of harm or adverse impact caused by the AI system?
2. How has the organization implemented mechanisms to provide information to end-users or third parties about opportunities for redress?[7]

RESPONSIBLE AI

What Is Responsible AI?

Responsible AI is a set of principles that help guide the design, development, deployment, and use of AI—building trust in AI solutions that have the potential to empower organizations and their stakeholders.

Responsible AI involves the consideration of the broader societal impact of AI systems and the measures required to align these technologies with stakeholder values, legal standards, and ethical principles. Responsible AI aims to embed such ethical principles into AI applications and workflows to mitigate risks and negative outcomes associated with the use of AI while maximizing positive outcomes.

AI must be trustworthy, and for stakeholders to trust AI, it must be transparent.[8]

In the end, responsible AI is a framework that addresses the potential risks and consequences of AI, including biases, privacy concerns, security threats, and unintended harm while ensuring AI benefits individuals and organizations alike.

Responsible AI Framework

- Ethical Considerations
  AI can perpetuate or amplify biases, making it crucial to ensure systems operate fairly and without discrimination.
- Transparency
  Lack of explainability in AI decisions can lead to loss of trust, especially in high-stakes industries.
- Accountability
  Organizations must ensure they can track and address issues arising from AI use.
- Compliance
  Adherence to laws, regulations, and ethical guidelines helps mitigate legal and reputational risks.
- Risk Mitigation
  Understanding and mitigating risks (e.g., privacy breaches, algorithmic bias, system malfunctions) is vital.

- Sustainability
  AI should be used responsibly to contribute positively to the environment and society.

Auditing Responsible AI Principles – General Questions

Accountability and Risk Management

1. How are incidents related to AI systems identified, documented, and addressed?
2. Describe the process in place for addressing negative consequences or errors.
3. What processes are in place to evaluate the risks associated with AI and ensure accountability for decisions made by AI systems?
4. Who is responsible for the outcomes of the AI tool's decisions?

Bias and Fairness

1. How does the organization ensure that its AI systems provide equitable outcomes for all user groups?
2. How is the AI tool tested for bias in its decisions?
3. What measures are in place to ensure fairness across different demographic groups?
4. What steps are taken to identify, measure, and mitigate biases in AI models, data, and outputs?

Collaboration, Consultation, Continuous Monitoring, and Improvement

1. Have stakeholders from diverse backgrounds been consulted about the AI tool's implementation?
2. How are AI systems monitored post-deployment to ensure they perform as intended over time?
3. Is there a plan for ongoing collaboration with relevant experts and stakeholders?
4. What processes exist for updating or retraining AI models to address changing conditions or new data?

Data Collection and Usage

1. What types of data will the AI tool require for training and operation?
2. Is the data being used ethically and with proper permissions?

Data Integrity, Security, and Privacy

1. How does the organization ensure the quality, integrity, and appropriateness of data used to train AI models?
2. How is sensitive user data handled and protected?

3. What safeguards are in place to protect user privacy and comply with relevant data protection regulations (e.g., GDPR, CCPA)?

Ethical Alignment and Framework

1. How does the organization ensure AI systems align with its ethical values, standards, and societal responsibilities?
2. What ethical guidelines and principles guide the development and deployment of the AI tool?
3. What mechanisms are used to identify and address ethical risks in AI system development?

Governance, Oversight, and Human Intervention

1. Can humans override the AI's decisions when necessary? How can this be performed? How is this procedure documented? Communicated to approved personnel?
2. How are responsibilities for AI oversight defined across stakeholders (e.g., data scientists, IT, legal, compliance)?
3. How does the AI tool handle cases it is uncertain about?
4. What governance framework is in place to oversee AI development, deployment, and monitoring?

Long-Term Effects

1. How will the AI tool adapt to changing data and circumstances over time?
2. What are the potential long-term effects of using this AI tool in the organization's operations?

Regulatory Compliance

1. How does the organization ensure compliance with relevant regulations, standards, and guidelines for AI systems?
2. What steps are taken to monitor and address evolving regulatory requirements for AI usage?

Security Controls

1. What security controls are in place to protect AI systems from tampering, adversarial attacks, or unauthorized access?
2. How is the organization ensuring the resilience of AI systems against cyber threats?

Societal Impact

1. Have potential negative societal impacts of the AI tool been identified and addressed?
2. Could the AI tool exacerbate existing social issues?

Stakeholder Engagement

1. How does the organization engage with stakeholders (e.g., employees, customers, regulators) to gather feedback on AI usage?

2. What steps are taken to address concerns or grievances raised by stakeholders about AI systems?

Transparency and Explainability

1. Can the AI's decision-making process be easily understood and explained to non-technical users responsible for implementing and using the AI tool?
2. How does the AI tool make decisions?
3. How does the organization ensure that AI decisions are explainable and understandable to stakeholders?
4. What measures are in place to communicate AI logic and decisions to end-users, especially in critical use cases?

NOTES

1 OECD (2019), Artificial Intelligence in Society, OECD Publishing, Paris, https://doi.org/10.1787/eedfee77-en.

2 Figure 7.1, drawn by author based on material from National Institute of Standards and Technology, January 2023, "Artificial Intelligence Risk Management Framework (AI RMF 1.0)," www.nist.gov/itl/ai-risk-management-framework, https://doi.org/10.6028/NIST.AI.100–1.

3 National Institute of Standards and Technology, January 2023, "Artificial Intelligence Risk Management Framework (AI RMF 1.0)," www.nist.gov/itl/ai-risk-management-framework, https://doi.org/10.6028/NIST.AI.100–1.

4 AI HLEG, "EU's Ethics Guidelines for Trustworthy Artificial Intelligence," High-Level Expert Group on Artificial Intelligence, European Commission, https://ec.europa.eu/newsroom/dae/document.cfm?doc_id=60419.

5 Juan, January 11, 2023, "Algorithm Audit for a Trustworthy AI," PickleTech, https://pickletech.eu/blog-audit/#:~:text=Final%20words,and%20continuously%20assessed%20during%20deployment.

6 Crockett, K., Colyer, E., Gerber, L., and Latham, A., August 2023, "Building Trustworthy AI Solutions: A Case for Practical Solutions for Small Businesses," in *IEEE Transactions on Artificial Intelligence*, vol. 4, no. 04, pp. 778–791, doi: 10.1109/TAI.2021.3137091. https://e-space.mmu.ac.uk/629069/1/TAI3137091%20corrected%20final%20version%20colour%20photos.pdf.

7 The AI-ELEG audit questions were modified into open-ended internal audit questions by the author, based on AI HLEG, "EU's Ethics Guidelines for Trustworthy Artificial Intelligence," High-Level Expert Group on Artificial Intelligence, European Commission, https://ec.europa.eu/newsroom/dae/document.cfm?doc_id=60419.

8 Stryker, C., February 6, 2024, "What is responsible AI?" IBM, www.ibm.com/topics/responsible-ai#:~:text=Published%3A%206%20February%202024,empower%20organizations%20and%20their%20stakeholders.

Chapter 8

AI Systems Decision-Making

Auditing Fairness

As artificial intelligence (AI) systems continue to evolve and impact the daily lives of global citizens, sometimes even without our knowledge, the potential for biases and, therefore, potential discrimination within the actions taken and the responses provided by these systems grows exponentially.

One cannot discuss fairness in decision-making without at least a rudimentary discussion of a key element that factors into making fair decisions....that being the ability to be unbiased in the decision-making process.

> A purely technical approach is insufficient to understand and mitigate biases. It perpetuates the misleading notion that ML systems can achieve "fairness" or be "unbiased".[1]

Bias is defined as a systematic error in decision-making processes that results in unfair outcomes. In the context of AI, bias can arise from various sources, including data collection, algorithm design, and human interpretation. Machine-learning models (MLMs), a type of AI system, can learn and replicate patterns of bias present in the data used to train them, resulting in unfair or discriminatory outcomes. It is important to identify and address bias in AI to ensure that these systems are fair and equitable for all users.[2]

SOURCES OF BIAS IN AI

Bias in AI systems can be introduced via several intentional and unintentional sources such as:

- Data Collection
  Bias can originate from the data used to train AI models, such as datasets that are incomplete, unrepresentative, or skewed toward certain groups or outcomes.

 DOI: 10.1201/9781003646501-10

- Labeling and Annotation

  This refers to the process of adding metadata to data that helps an AI system learn from it (the metadata). Human bias may be introduced during the labeling or annotation process, where subjective judgments influence the data's interpretation.

  For example:

  Labeling: Assigning specific tags or categories to data points (e.g., tagging an image as "cat" or "dog"). This provides the ground truth that supervised learning models use for training.

  Annotation: Adding detailed information or context to raw data, such as marking objects in an image with bounding boxes, highlighting specific regions in a medical scan, or transcribing audio files.

  The bias in this step can arise if the human annotators unintentionally or intentionally mislabel data, rely on personal assumptions, or fail to consider diverse perspectives. This can skew the AI model's understanding and perpetuate biased or incomplete learning.
- Historical Inequities

  Training data that reflects historical inequalities or systemic discrimination can perpetuate or amplify biases in AI systems.
- Feature Selection

  The choice of features used in a model can inadvertently include proxies for sensitive attributes, leading to unintended bias.

  Feature selection refers to the process of identifying and selecting the most relevant input variables (features) from a dataset to use in training a MLM. These features are used by the algorithm to make predictions or classifications.

  Bias can arise during feature selection in the following ways:
  - Inclusion of Proxy Variables: A feature that appears neutral may indirectly correlate with sensitive attributes (e.g., ZIP codes acting as proxies for race or income level), unintentionally introducing bias.
  - Omission of Relevant Features: Excluding features that are crucial for making fair and accurate predictions can lead to biased outcomes, as the model might rely on less appropriate features.
  - Human Subjectivity: The process of selecting features often involves human judgment, which can introduce unintentional bias based on assumptions or preconceptions.
  - Data Imbalance: If certain features are underrepresented or missing in the dataset for specific groups, the model may not learn to treat those groups fairly.
- Algorithm Design

  Bias may be introduced through the design of algorithms that prioritize certain outcomes or inadvertently treat groups differently.

- Training Processes
  Models trained to optimize specific objectives, like accuracy, may sacrifice fairness or amplify disparities.
- Deployment Context
  Bias can emerge when models are deployed in contexts that differ significantly from the conditions under which they were trained.
  Example: An AI model was trained to predict auto insurance risk based on driving data from customers in temperate climates (mild winters, little snow). The model is later deployed in northern regions with heavy snowfall and icy roads, where driving behavior and accident patterns differ greatly. The model misinterprets cautious winter driving as risky behavior, resulting in unfair premium increases for safe drivers who adjust their driving habits to seasonal conditions.
  This mismatch between the model's training context and its deployment environment can lead to biased outcomes. It is important to ensure that AI models are tested and adjusted for the specific contexts in which they will be used.
- Feedback Loops
  Continuous use of biased models in real-world scenarios can create feedback loops, reinforcing existing biases over time.
- User Interactions
  Bias can be introduced through the way users interact with AI systems, particularly if the system adapts to user behavior in a way that reflects societal biases.
- Cultural and Social Assumptions
  Implicit cultural norms and assumptions embedded in the development process can influence AI behavior in biased ways.

TYPES OF BIAS IN AI

AI is dependent on a network of algorithms. These algorithms are essentially rules for how a machine operates on its own or with limited human assistance, providing instructions about how to learn from data and make decisions. Algorithms do so by automating these tasks at a robust scale and speed.

Algorithms used in AI systems may also produce bias due to how the data are collected or analyzed. Data samples can be incomplete, contain irrelevant information, or reflect historical prejudices. The algorithms can also mirror and compound even small biases from humans who train or supervise them.[3]

Table 8.1 summarizes the types of bias that may be found in AI systems.

Table 8.1 Types of Bias in AI Systems

Bias Risk	*Explanation and Example*
Aggregation Bias	Bias from applying a single model to diverse groups with different needs. Arises during model construction where distinct populations are inappropriately combined. Example: A healthcare algorithm assuming uniform disease symptoms across populations may misdiagnose minority groups.
Algorithmic Bias	Introduced during algorithm design or optimization. Example: Prioritizing accuracy over fairness can lead to unequal outcomes for different demographic groups.
Cultural Bias	Arises from cultural assumptions embedded in AI systems. Example: A voice assistant trained only on Western accents struggles to understand non-Western accents.
Deployment Bias	Bias resulting from mismatched training and deployment contexts. Example: An AI model trained in urban areas may perform poorly in rural settings due to contextual differences.
Evaluation Bias	Occurs when the metrics or benchmarks used to assess an AI system's performance are not representative or inclusive of all user groups. Example: A facial recognition system evaluated only on a dataset of light-skinned faces may show high accuracy during testing but fail to perform well on darker-skinned individuals in real-world applications.
Feedback Loop Bias	Bias created when AI systems reinforce their own outputs over time. Example: A content recommendation system continuously promotes specific types of media, reducing diversity.
Historical Bias	Bias inherent in historical data used to train AI models, reflecting societal inequities. Example: A hiring model trained on historical data may prefer male candidates due to past gender disparities in hiring.
Measurement Bias	Bias arising from how data is measured or labeled. Example: Inconsistent labeling of medical records could lead to biased diagnostic AI systems.
Omitted Variable Biasw	Bias due to exclusion of relevant variables in the model. Example: A credit-scoring system ignoring socioeconomic context may disadvantage low-income individuals.
Representation Bias	Representation bias occurs when certain groups are underrepresented or overrepresented in training data, leading to biased outcomes. Example: A voice assistant trained primarily on male voices may struggle to accurately interpret or respond to female voices, resulting in unequal performance and user dissatisfaction.
Selection Bias	Bias introduced due to unrepresentative training data. Example: A facial recognition system trained mostly on light-skinned faces misidentifies darker-skinned individuals.
Societal Bias	Bias stemming from societal norms and values embedded in data. Example: Language models reflecting stereotypes present in online content.

TYPES OF RISK FROM BIAS IN AI SYSTEMS

Table 8.2 identifies the type of risks organizations may face due to bias in AI systems.

Table 8.2 Risk from Bias in AI Systems

Risk	*Explanation and Example*
Economic Consequences	Bias in AI systems can cause financial losses due to inefficiencies, reduced customer base, or penalties. Example: A retail AI pricing algorithm unintentionally overcharges certain demographics, driving customers away and decreasing revenue.
Employee Dissatisfaction	Internal bias in AI systems can lead to employee frustration, dissatisfaction, and attrition. Example: A biased internal performance review system unfairly evaluating employees from underrepresented groups.
Ethical Violations	AI bias can cause organizations to act unethically, leading to societal backlash and moral concerns. Example: A biased school admissions algorithm prioritizes students from affluent neighborhoods (indicated by applicants' home zip code), exacerbating educational inequality and disadvantaging students from lower-income areas.
Hindered Innovation	Bias may limit the effectiveness of AI solutions and hinder the development of innovative technologies. Example: A biased dataset restricting the ability to explore diverse use cases for an AI tool.
Legal Consequences	Biased AI systems can lead to violations of laws and regulations, resulting in lawsuits and penalties. Example: A credit-scoring AI system found to unfairly reject loan applications for minority groups, leading to lawsuits under anti-discrimination laws.
Liability Risk	Organizations may face liability for harm caused by biased AI decisions. Example: A biased facial recognition system used in retail security falsely identifies innocent individuals as shoplifters, leading to reputational damage and legal action against the company.
Loss of Reputation	Organizations risk damaging their brand and public image if their AI systems exhibit bias or unfairness. Example: A biased recruitment tool publicly criticized for discriminating against women, leading to public backlash and a loss of customer trust.
Market Competitiveness Risk	Bias can harm competitiveness by alienating key customer segments. Example: A biased e-commerce recommendation system frequently excludes products targeted toward minority-owned businesses, reducing diversity and alienating socially conscious customers.

(*Continued*)

Table 8.2 (Continued)

Risk	*Explanation and Example*
Operational Inefficiency	Bias in AI systems can lead to incorrect decisions, reducing system effectiveness and operational efficiency. Example: A biased customer support bot failing to provide relevant solutions to specific demographics.
Reduced User Trust	Bias can erode user confidence and trust in an organization's AI systems and overall brand. Example: A biased hiring algorithm leading candidates to perceive the organization as discriminatory.
Regulatory Non-Compliance	Bias in AI systems can result in violations of regulatory requirements, resulting in penalties. Example: An AI system failing to adhere to fairness standards outlined in anti-discrimination laws.
Social Harm and Inequality	Biased AI systems may perpetuate or exacerbate societal inequities, causing broader harm. Example: A biased predictive policing system disproportionately targeting minority communities.

While fairness and bias are closely related concepts, they differ in important ways, even in the context of AI systems.

DEFINING FAIRNESS

The meaning of the term "fairness," in its contemporary English language usage, has dozens of interpretations. They include a range of related but distinctive ideas, such as equity, consistency, non-discrimination, impartiality, justice, equality, honesty, and reasonableness.[4]

When asked to define fairness, or what it means to act fairly, given admittedly dozens of interpretations, whose meaning may differ among individuals, most respondents' replies, however, will identify and include several core elements that have come to embody the general concept or understanding of fairness or what it means to be fair. These elements are:

1. Equality

   Ensuring that all individuals are treated the same, with equal access to opportunities, resources, and rights, regardless of their personal or social characteristics.
2. Equity

   Recognizing that equal treatment may not always result in fair outcomes and addressing individual needs to achieve a balanced and just outcome.
3. Justice

   Upholding moral and legal principles to ensure that decisions and actions are based on fairness, honesty, and respect for rights.

4. Impartiality
   Making decisions without favoritism, prejudice, or bias, and ensuring a neutral stance in resolving conflicts or disputes.
5. Proportionality
   Ensuring that actions and decisions are reasonable and appropriate to the circumstances, avoiding excessive or disproportionate measures.
6. Transparency
   Ensuring openness and clarity in decision-making processes, allowing stakeholders to understand how and why decisions are made.
7. Accountability
   Holding individuals, institutions, or systems responsible for their actions, ensuring they are answerable for outcomes.
8. Respect for Diversity
   Recognizing and valuing differences among individuals or groups, fostering inclusion and understanding to prevent discrimination.
9. Consistency
   Applying rules, standards, or principles uniformly across cases or situations to avoid arbitrary treatment.
10. Reciprocity
    Ensuring mutual respect and consideration in interactions, where fairness is upheld by balancing the needs and interests of all parties involved.

Based on these central elements and the notion that each person possesses an intrinsic and equal moral worth and that everyone deserves equal respect and concern, Leslie et al. put forth the following definition of fairness:

> Fairness has to do with the moral duty to treat others as moral equals and to secure the membership of all in a 'moral community' where every person can regard themself as having equal value. Wrongful discrimination occurs when decisions, actions, institutional dynamics, or social structures do not respect the equal moral standing of individual persons.[5]

WHAT IS FAIRNESS IN AI?

> Fairness is a crucial element in AI, for without it, how are we to construct responsible and ethical AI systems?

We trust ML systems, expecting them to be "fair" and not discriminatory (especially not to discriminate against groups that are legally protected, such as by race and gender). But it is not that simple. Rarely are different definitions and notions of fairness considered at the start of developing an ML system.

Even if different definitions or approaches are considered, there is not necessarily a "right" answer for a particular AI system. Also, there are various actors involved in the ML process (from dataset development through the development of algorithms and the use of AI systems) that might have different understandings and interpretations of fairness.[6]

The concern of fairness in AI centers on the risk for algorithms to display bias, resulting in discriminatory outcome. Bias may unintentionally arise during the data collection and training processes (from biases embedded in the training data), or in the way the systems were designed, often resulting in AI systems reflecting historical prejudices and perpetuating existing inequalities.

Addressing the issue of fairness is critical for building trust in AI systems and ensuring the systems designed and placed into production serve our global society.

Why Is AI Fairness Important?

AI fairness is essential to prevent discrimination, promote diversity, and uphold ethical standards in technology.

Without fairness, AI systems may inadvertently amplify existing societal biases, leading to unfair treatment and negative consequences for certain individuals or communities. Prioritizing fairness fosters trust in AI applications, encourages widespread adoption, and supports the development of technology that benefits everyone.

Fairness in AI is not only a matter of social justice but also a legal and ethical obligation.[7]

TYPES OF AI FAIRNESS

The examination of varying types of fairness found in AI systems has been addressed by multiple authors/researchers (i.e., Emilio Ferrara et al.,[8] David Leslie et al.,[9] Zafar et al.,[10] and Kleinberg et al.[11]).

Table 8.3 provides a summarized list of AI fairness types, accompanied by a short description of each and an example of the application of "fairness."

ACHIEVING FAIRNESS IN AI SYSTEMS

Achieving fairness in AI systems first requires understanding and addressing potential biases in the data and algorithms that power AI systems. Bias creeps into both the data used to train MLMs and intentionally or not in the programming designed by those developing AI systems.

Addressing bias inherent in AI systems may be accomplished by:

- Ensuring that the datasets used to train AI systems are diverse and representative of the population they serve. This involves identifying and mitigating biases in data collection and preprocessing stages to prevent skewed outcomes.
- Using fairness metrics to evaluate AI models and applying algorithmic fairness approaches, such as reweighing, to adjust models that show discriminatory biases.

Table 8.3 AI Fairness Types

AI Fairness Type	*Definition*	*Example*
Application Fairness	Ensures that the objectives and policy choices driving AI systems align with fairness and equity considerations, and do not exacerbate inequities or systemic injustices.	An AI system designed for job recruitment ensures that its policies are aligned with community equity concerns and does not disproportionately exclude underrepresented groups.
Causal Fairness	Focuses on identifying and understanding the causal relationships in data to ensure fairness by addressing underlying causes of biases in outcomes.	An AI model identifies that differences in medical diagnoses are due to historical inequities in healthcare access, adjusting to ensure fairness in outcomes.
Counterfactual Fairness	Requires that an AI model produces the same decision for an individual in both the real world and a hypothetical counterfactual world where sensitive attributes are changed.	A loan approval system would approve or reject an applicant in the same way, regardless of their ability to repay, in a hypothetical world where the applicant's financial status is different.
Data Fairness	Addresses the fairness of data collection, handling, and usage, ensuring that datasets are representative, accurate, and free of systemic biases.	A facial recognition system ensures its training data includes diverse demographics to avoid misidentifications based on race or gender.
Ecosystem Fairness	Involves fairness in the broader ecosystem where AI systems are deployed, ensuring equitable influence and outcomes in economic, legal, and social structures.	A social media algorithm ensures that its deployment does not amplify discriminatory content or reinforce existing power imbalances in marginalized communities.
Group Fairness	Ensures fairness across groups by balancing outcomes and error rates among different demographic or protected groups.	A university admissions model ensures that applicants from different socioeconomic backgrounds have equal chances based on their qualifications.
Individual Fairness	Focuses on treating individuals with similar relevant characteristics in a similar way, ensuring fairness on a per-person basis.	A credit-scoring system ensures that two applicants with identical financial histories receive the same credit decision, regardless of their demographic attributes.

(*Continued*)

Table 8.3 (Continued)

AI Fairness Type	*Definition*	*Example*
Metric-Based Fairness	Uses mathematical metrics to measure and enforce fairness in AI systems, operationalizing principles like equality of outcomes or error rates.	An AI hiring tool measures fairness by ensuring that positive hiring decisions are equally distributed among gender groups.
Model Design and Development Fairness	Ensures fairness in the design, development, and training of AI models by avoiding discriminatory features, proxies, or historical biases.	An AI healthcare system avoids using healthcare costs as a proxy for health needs, addressing disparities caused by socioeconomic factors.
Procedural Fairness	Focuses on fairness ensuring that the process used to make decisions is fair and transparent.	An AI model designed with a transparent decision-making process.
System Implementation Fairness	Addresses fairness in how AI systems are implemented, ensuring they are used responsibly with consideration for their impacts on affected individuals.	An AI-powered customer service chatbot is deployed with regular reviews to ensure it does not generate biased or offensive responses.

- Regularly monitoring AI systems for biased outcomes and recalibrating models as necessary is essential to maintain fairness over time, especially as societal norms and values evolve.[12]

Designing fairness in AI systems requires users to understand what goes on behind the scenes, how AI analyzes the data provided, decides among potential alternatives, reaches conclusions, and presents decisions. Fairness requires transparency!

Transparency in AI systems refers to making the processes, decision-making logic, and underlying data of AI models understandable and accessible to stakeholders. It ensures that users, developers, and regulators can comprehend how the system functions, including its design, training, and potential biases, promoting trust and accountability.

In addition to building trust, AI transparency fosters knowledge-sharing and collaboration across the entire AI ecosystem, contributing to advancements in AI development. And by being transparent by default, organizations can focus more on using AI technologies to achieve business goals – and worry less about AI reliability.[13]

The following are fundamental recommendations for achieving transparency in AI, and ultimately achieving fairness:

1. Implement Explainable AI (XAI) Techniques
   Use methods and models that provide clear and interpretable explanations of AI decisions, enabling stakeholders to understand the reasoning behind the outcomes.
2. Document the AI Lifecycle
   Maintain comprehensive documentation of the AI development process, including data sources, preprocessing steps, model architecture, and evaluation metrics, to ensure traceability and accountability.
3. Adopt Transparent Algorithms
   When and where feasible and applicable, design and use inherently interpretable algorithms, such as decision trees or linear models, over opaque models like deep neural networks.
4. Provide Stakeholder Communication
   Create user-friendly summaries or visualizations of how the AI system works and how decisions are made, tailored to different stakeholder groups, including non-technical audiences.
5. Establish Auditing and Monitoring Mechanisms
   Conduct regular audits of AI systems to detect biases or unexpected behavior and implement monitoring tools that track system performance and decision-making over time.

As previously defined, fairness in AI systems refers to ensuring that algorithmic decisions are free from bias and do not result in unjust or discriminatory outcomes for individuals or groups. It involves designing, training, and deploying AI models in a manner that treats all users equitably, considering the social, cultural, and legal contexts of their application.

Achieving fairness in AI systems requires developers to design and users to require AI systems to embrace globally accepted ethical principles and comply with established and evolving global regulations and standards for the moral development of AI.

The most significant global regulations and standards for the ethical development of AI that should guide organizations in the current and ongoing design and development of AI systems include:

1. OECD AI Principles (Organisation for Economic Co-operation and Development): Established in 2019, these principles emphasize transparency, accountability, and human-centered values, providing a global framework for responsible AI development.
2. EU Artificial Intelligence Act: A proposed regulatory framework by the European Union, focusing on risk-based AI governance, ensuring safety, transparency, and non-discrimination in AI applications.
3. UNESCO Recommendation on the Ethics of Artificial Intelligence: Adopted in 2021, this provides an international standard for ethical

AI, emphasizing human rights, sustainability, and fairness in AI systems.
4. IEEE Global Initiative on Ethics of Autonomous and Intelligent Systems: This initiative provides comprehensive ethical guidelines, including the "Ethically Aligned Design" framework, promoting transparency, accountability, and fairness in AI development.
5. ISO/IEC JTC 1/SC 42 – Artificial Intelligence: An international standardization initiative focusing on AI governance, addressing ethical, technical, and operational aspects to promote global consistency in AI practices.

While not enforceable, significant frameworks are designed to guide organizations in responsible (fair) development and use of AI.

These include but, are not limited to:

- AI Ethics Principles from the United States Department of Justice (DOJ).
- AI Now Institute's Algorithmic Accountability Framework.
- Blueprint for an AI Bill of Rights (White House Office of Science and Technology Policy).
- Defense Innovation Board's Ethical Principles for AI.
- Ethically Aligned Design (IEEE Global Initiative on Ethics of Autonomous and Intelligent Systems).
- Federal Trade Commission (FTC) Guidance on AI.
- National Institute of Standards and Technology (NIST) AI Risk Management Framework.
- Singapore Model AI Governance Framework.
- The AI Ethics Guidelines by the European Commission's High-Level Expert Group on AI.
- The Asilomar AI Principles.
- The Montreal Declaration for Responsible AI.
- The Partnership on AI (PAI) Framework.
- The Toronto Declaration.
- UN Guiding Principles on Business and Human Rights Applied to AI.
- World Economic Forum (WEF) Toolkit for Responsible AI.

Auditors and risk assessors should review the above, developing and incorporating key audit/assessment questions based on the content and directives of these standards and frameworks into their respective audit programs.

Achieving fairness in AI must ultimately be part of the organization's culture, supported by executive management, required by users and stakeholders, and designed and incorporated into AI systems by data scientists and developers.

It is important to note that attaining fairness in AI is not a one-size-fits-all solution and requires careful consideration of the context and stakeholders involved.[14]

ASSESSING FAIRNESS: QUANTITATIVE AND QUALITATIVE TOOLS

The following quantitative tools are used by engineers and data scientists to examine, report, and mitigate discrimination and bias in ML models.

Auditors and risk assessors can benefit from exploring the functional capabilities of these tools, designed to evaluate and mitigate bias in MLMs and AI systems when developing internal control questions to assess an AI system's fairness approach.

Quantitative Tools

- IBM's AI Fairness 360 Toolkit

  An open-source library designed to help developers detect and mitigate bias in MLMs. It provides a comprehensive set of metrics to evaluate fairness and includes algorithms to reduce bias, catering to multiple-use cases across industries. (https://aif360.res.ibm.com)
- What-If Tool

  Developed by Google, a user-friendly visualization tool that allows developers and analysts to experiment with MLMs. By enabling "what-if" scenarios and providing insights into individual and aggregated predictions, it helps explore model behavior and potential biases without requiring code changes. (https://pair-code.github.io/what-if-tool)
- Microsoft's fairlean.py

  An open-source Python library focused on improving fairness in AI systems by addressing and measuring biases. It includes tools for assessing the fairness of MLMs and offers algorithms to balance performance across different demographic groups. (https://techcommunity.microsoft.com/t5/educator-developer-blog/fairlearn-a-python-package-to-assess-ai-system-s-fairness/ba-p/1402950)
- Facebook's "Fairness Flow"
- An internal AI tool that automates the process of evaluating machine learning models for fairness. It integrates into the company's broader AI workflow, enabling teams to identify and address potential biases in models used for large-scale applications. [https://ai.meta.com/blog/how-were-using-fairness-flow-to-help-build-ai-that-works-better-for-everyone]

- Note: While this tool evaluates MLMs for fairness, access to the tool is restricted and not open source to external developers or other interested parties (i.e., auditors).

Qualitative Tools

- Fairkit-learn
  A toolkit that assists data scientists in evaluating and understanding fairness trade-offs in machine learning models, integrating with existing tools to provide insights into model performance concerning fairness metrics. This tool was proposed by Johnson, B., Bartola, J., Angell, Rico, et al. in their paper "Fairkit, Fairkit, on the Wall, Who's the Fairest of Them All? Supporting Data Scientists in Training Fair Models" (2020). [https://arxiv.org/pdf/2012.09951]
- FairCompass
  A human-in-the-loop visual analytics system designed to operationalize fairness auditing in machine learning models, enabling users to explore and mitigate biases through an interactive interface. This tool was proposed by Liu, J., Chen, H., et al., in their paper "FairCompass: Operationalising Fairness in Machine Learning" (2023). [https://arxiv.org/abs/2312.16726?utm_source=chatgpt.com]

AUDITING FAIRNESS IN AI SYSTEMS

The following set of 25 questions serves as an introductory resource, a starting point for auditors and risk assessors to evaluate the concept of fairness in AI decision-making systems. These questions encourage a deep examination of fairness across the AI lifecycle, from data collection and design to deployment and monitoring.

By addressing structural inequalities, mitigating biases, and ensuring equitable outcomes, these questions aim to help organizations assess the alignment of AI systems with ethical standards and societal expectations. The goal is to provide a structured framework for the assessment of the AI tool's fairness by examining its data, development processes, decision-making transparency, impact on diverse groups, and ongoing monitoring practices.

Auditing Fairness in AI Systems - Audit Questions

1. What methods have been employed to identify structural inequalities and institutionalized patterns of discrimination that the AI system could inadvertently replicate or exacerbate?
2. How does the AI system's design explicitly align with broader social justice goals to rectify inequities and promote equitable outcomes?

3. What strategies are in place to identify and mitigate historical biases embedded in the training data, and how are these efforts evaluated for effectiveness?
4. How have marginalized communities and other impacted stakeholders been included in decision-making to ensure the AI system addresses their needs and concerns?
5. What measures are used to assess whether the AI system's outputs equitably benefit all demographic groups, especially those historically disadvantaged?
6. In what ways does the AI system support systemic change by addressing root causes of inequality rather than merely treating symptoms?
7. How does the system account for the compounded impact of intersecting identities, such as race, gender, and socioeconomic status, on affected individuals and groups?
8. How does the AI system comply with ethical standards and legal requirements designed to promote fairness and prevent discrimination?
9. What specific methods were used to balance fairness, efficiency, and other priorities during system development, and how are these decisions monitored over time for continued relevance?
10. What mechanisms are in place to continuously monitor and improve the system's performance in addressing structural inequalities and preventing discrimination over its lifecycle?
11. How were the perspectives of marginalized communities integrated into the AI system's lifecycle, and what mechanisms ensured their meaningful participation in decision-making?
12. Beyond identifying potential biases, what measures are taken to validate the effectiveness of data bias mitigation strategies across diverse user groups?
13. What fairness metrics have been selected for evaluating this AI system, and why were these specific metrics chosen over others for this context?
14. How does the AI system assess its potential impact on protected or vulnerable groups, and what measures have been implemented to mitigate negative impacts?
15. How are the system's decision-making processes and potential fairness concerns communicated to stakeholders, including non-technical users?
16. Who is responsible for addressing fairness-related issues, and what accountability mechanisms ensure timely remediation of these issues?
17. How does the AI system adapt its fairness considerations to align with the specific social, cultural, and legal context in which it is deployed?
18. How does the AI system proactively ensure equitable outcomes for vulnerable populations in both the design and deployment phases, and how are these outcomes measured?

19. What specific accountability structures and performance indicators are in place to address and resolve fairness-related issues in the system?
20. How have the fairness-related assumptions, decisions, and mitigation efforts been documented, and what steps have been taken to reflect on and improve these efforts?
21. What specific aspects of the AI system's design, development, or deployment could contribute to outcomes that may be perceived as discriminatory harassment by impacted individuals?
22. How are policies and decision-making processes integrated to ensure the system avoids fostering environments or outcomes that could lead to harassment?
23. In what ways might the design or operational processes of the AI system lead to disproportionately negative outcomes for individuals from protected groups?
24. How could indirect biases in the AI system's data, design, or decision-making processes result in discriminatory outcomes for individuals from protected groups?
25. What measures have been implemented to identify and mitigate potential sources of indirect discrimination in the system?

NOTES

1 Smith, G., Kohli, N., and Rustagi, I., 2020, "What does "fairness" mean for machine learning systems?" An accompanying resource of Mitigating Bias in Artificial Intelligence: An Equity Fluent Leadership Playbook of the Center for Equity, Gender & Leadership (EGAL) at Berkeley Haas, https://haas.berkeley.edu/wp-content/uploads/What-is-fairness_-EGAL2.pdf.
2 Ferrara, E., December 26, 2023, "Fairness and Bias in Artificial Intelligence: A Brief Survey of Sources, Impacts, and Mitigation Strategies," Sci Journal, https://doi.org/10.3390/sci6010003.
3 Pontón-Núñez, A., April 8, 2024, "Addressing the Risks and Harms of Artificial Intelligence by Leveraging Capital," Nonprofit Quarterly, https://nonprofitquarterly.org/addressing-the-risks-and-harms-of-artificial-intelligence-by-leveraging-capital.
4 Leslie, D., Rincón, C., Briggs, M., Perini, A., Jayadeva, S., Borda, A., Bennett, SJ. Burr, C., Aitken, M., Katell, M., Fischer, C., Wong, J., and Kherroubi Garcia, I., 2023, "AI Fairness in Practice," The Alan Turing Institute, www.turing.ac.uk/sites/default/files/2023-12/aieg-ati-fairness_1.pdf.
5 Leslie, D., Rincón, C., Briggs, M., Perini, A., Jayadeva, S., Borda, A., Bennett, SJ. Burr, C., Aitken, M., Katell, M., Fischer, C., Wong, J., and Kherroubi Garcia, I., 2023, "AI Fairness in Practice," The Alan Turing Institute, www.turing.ac.uk/sites/default/files/2023-12/aieg-ati-fairness_1.pdf.
6 Smith, G., Kohli, N., and Rustagi, I., 2020, "What does "fairness" mean for machine learning systems?" An accompanying resource of Mitigating Bias in Artificial Intelligence: An Equity Fluent Leadership Playbook of the Center

for Equity, Gender & Leadership (EGAL) at Berkeley Haas, https://haas.berkeley.edu/wp-content/uploads/What-is-fairness_-EGAL2.pdf, individual tool descriptions summarized independently by the author.

7 (n.a.), January 26, 2024, "AI Fairness," Clan, https://clanx.ai/glossary/ai-fairness.

8 Ferrara, E., December 26, 2023, "Fairness and Bias in Artificial Intelligence: A Brief Survey of Sources, Impacts, and Mitigation Strategies," Sci Journal, https://doi.org/10.3390/sci6010003.

9 Leslie, D., Rincón, C., Briggs, M., Perini, A., Jayadeva, S., Borda, A., Bennett, SJ. Burr, C., Aitken, M., Katell, M., Fischer, C., Wong, J., and Kherroubi Garcia, I., 2023, "AI Fairness in Practice," The Alan Turing Institute, www.turing.ac.uk/sites/default/files/2023-12/aieg-ati-fairness_1.pdf.

10 Zafar, M. B., Valera, I., Gomez Rodriguez, M., and Gummadi, K. P., April 3, 2017, "Fairness beyond Disparate Treatment & Disparate Impact: Learning Classification without Disparate Mistreatment. In *Proceedings of the 26th International Conference on World Wide Web* 1171–1180 pages, https://arxiv.org/pdf/1610.08452.

11 Kleinberg, J., Ludwig, J., Mullainathan, S., and Rambachan, A. May 2018, "Algorithmic fairness," In *AEA papers and proceedings* (Vol. 108, pp. 22–27), www.cs.cornell.edu/home/kleinber/aer18-fairness.pdf.

12 Crosby, P., March 13, 2024, "How do I Ensure Fairness and Transparency in AI?" The League of Analysts Inc., https://theuncommonleague.com/articles/how-do-i-ensure-fairness-and-transparency-in-ai#:~:text=Regularly%20monitoring%20AI%20systems%20for%20biased%20outcomes,and%20values%20evolve.%20Ensuring%20Transparency%20in%20AI.

13 Joinker,A.,Gomstyn,A.,andMcGrath,A.,September6,2024,"WhatisAItransparency?" IBM, www.ibm.com/think/topics/ai-transparency#:~:text=AI%20creators%20can%20achieve%20transparent,evaluation%20and%20validation%20and%20more.

14 Ferrara, E., December 26, 2023, "Fairness and Bias in Artificial Intelligence: A Brief Survey of Sources, Impacts, and Mitigation Strategies," Sci Journal, https://doi.org/10.3390/sci6010003.

Chapter 9

Auditing the Reliability of Generative AI Tools

AUDITING THE RELIABILITY OF GENERATIVE AI

Auditing the reliability of generative AI tools is crucial for several reasons:

1. Trust and Credibility
   - Users need to trust that generative AI tools produce accurate, unbiased, and reliable outputs. An audit provides a systematic way to validate the integrity of the models and algorithms behind these tools, ensuring their results are consistent and dependable.
2. Accountability and Transparency
   - Audits help organizations be transparent about how generative AI tools operate, what data they use, and how decisions are made. This transparency is essential for building accountability, especially in sectors with stringent ethical and regulatory requirements.
3. Mitigation of Bias and Ethical Concerns
   - Many generative AI tools can inadvertently produce biased outputs based on the data they were trained on. Audits can detect and mitigate biases, ensuring that these tools make fair and inclusive decisions, particularly in sensitive applications like hiring, healthcare, or law.
4. Compliance and Regulatory Standards
   - With increasing regulations on AI and data usage, auditing ensures compliance with laws such as GDPR, AI Act, and other regional standards. It can protect organizations from legal risks and fines related to improper use of data or flawed AI operations.
5. Data Security and Privacy Assurance
   - Generative AI tools often work with large datasets, some of which may include sensitive information. Auditing helps ensure that the tools follow robust data security and privacy practices, protecting users and organizations from potential data breaches or misuse.
6. Performance Optimization
   - Audits evaluate the performance, efficiency, and accuracy of generative AI tools. Regular audits can identify areas for improvement,

DOI: 10.1201/9781003646501-11

leading to optimizations that make the tools more effective and resource-efficient.

7. Risk Management
    - Relying on generative AI tools without understanding their limitations and potential failure points poses significant risks. Auditing allows for the identification of such risks, enabling organizations to implement measures to minimize the impact of unexpected outputs or model failures.

By asking the following questions, auditors can assess the reliability of generative AI tools and evaluate their integration into the organization's workflows.

- **What Is the Track Record of Reliability?**
    1. What is the tool's historical performance in terms of reliability and stability?
    2. Are there any documented instances of the tool failing or producing inaccurate results? If yes, what were the causes and consequences?
- **Performance and Accuracy**
    1. How accurate are the tool's outputs compared with established benchmarks or human-generated results?
    2. What is the margin of error or uncertainty associated with the tool's predictions or recommendations?
    3. Can the tool provide evidence or examples of its accurate performance in various scenarios?
- **How is Reliability Defined and Measured?**
    1. How do you define reliability in the context of this AI tool?
    2. What metrics or Key Performance Indicators are used to measure and track the tool's reliability over time?
- **What Safety Measures Are in Place?**
    1. What safety mechanisms are implemented to prevent the tool from generating harmful or misleading outputs?
    2. How does the tool handle sensitive or confidential information to ensure data security and privacy?
- **What Quality Assurance Processes Are Followed?**
    1. What testing processes are used to ensure the tool's reliability before and after deployment?
    2. Are there systematic procedures in place to identify and address potential issues that might affect reliability?
- **How Does the Tool Handle Uncertainty?**
    1. How does the tool handle ambiguous or uncertain situations where it lacks sufficient data to provide a reliable output? How does it communicate uncertainty in its outputs?
    2. Is there a feedback loop in place to collect user feedback and use it to improve the tool's reliability?
    3. How frequently is the tool updated or retrained based on user feedback and changing requirements?

4. Can the tool recognize when it is unsure or lacks sufficient data to generate accurate results?
5. Is the tool transparent about its level of confidence in its predictions?

- **What Is the Training Data Quality?**
  1. What types of data were used to train the AI model? How diverse, representative, and up-to-date is the training data?
  2. Have efforts been made to mitigate biases and ensure fairness in the training data?
- **Data and Training**
  1. What is the source of the training data used to develop the AI model?
  2. How representative is the training data of the real-world situations the tool will encounter?
  3. Have efforts been made to address biases in the training data, and how is fairness ensured?
- **Can the Tool Adapt to Changing Conditions?**
  1. How does the tool handle situations where the operating conditions change or evolve over time?
  2. Is there a feedback loop for continuous improvement based on user interactions and new data?
- **How Does the Tool Handle Out-of-Distribution Inputs?**
  1. How does the tool react to inputs that are significantly different from the data it was trained on?
  2. Is there a risk of producing unreliable outputs when faced with inputs that fall outside its training distribution?
- **Adaptability and Generalization**
  1. How well does the tool perform when faced with inputs or situations that are slightly different from its training data?
  2. Can the tool adapt to new or evolving trends, contexts, or domains without significant loss of reliability?
- **What Redundancy or Backup Plans are in Place?**
  1. Are there backup systems, contingency plans, or redundancy strategies in place in case the tool experiences downtime or malfunctions?
  2. How quickly can the organization switch to an alternative solution if needed?
  3. How quickly can the organization switch to an alternative solution if the tool's reliability is compromised?
- **What Level of Human Oversight Is Required?**
  1. To what extent does the tool require human supervision, oversight, or validation of its outputs? How is this balance achieved?
  2. How does the tool balance automation and human intervention to ensure reliable results?
  3. Can the tool provide explanations or reasoning for its decisions to facilitate human understanding and validation?
- **How Is User Feedback Incorporated?**
  1. How does the organization gather feedback from users about the tool's reliability and performance?

2. Are there mechanisms in place to address user concerns and iteratively improve the tool?
3. Is there a feedback loop in place to collect user feedback and use it to improve the tool's reliability?
4. How frequently is the tool updated or retrained based on user feedback and changing requirements?

- **What Is the Communication Strategy during Failures?**
  1. How does the organization plan to communicate with users in the event of reliability issues or failures?
  2. What steps will be taken to maintain transparency and manage user expectations?
  3. What mechanisms are in place to detect and mitigate errors or failures in the tool's output?
  4. How quickly can the tool recover from errors, and what are the potential consequences of failures?
- **Security and Privacy**
  1. How does the tool handle sensitive data, and what measures are in place to ensure data security and user privacy?
  2. Has the tool undergone security audits or assessments to identify vulnerabilities?
- **Long-Term Reliability**
  1. How has the tool performed over an extended period? Are there any instances of degradation in performance?
  2. What guarantees or assurances can the tool provider offer regarding its reliability over time?
- **Regulatory and Compliance**
  1. Does the use of the tool comply with relevant industry regulations and standards?
  2. How does the tool provider address potential legal or ethical concerns related to reliability?
- **Communication and Transparency**
  1. How transparent is the tool's decision-making process, and can it provide insights into how it arrives at specific conclusions?
  2. How does the tool communicate its limitations and potential risks to users?

Chapter 10

Auditing the Ethics of AI

WHAT IS AI ETHICS?

Artificial intelligence (AI) ethics is a system of moral principles and techniques intended to inform the development and responsible use of AI technology. As AI has become integral to products and services, organizations are also starting to develop AI code of ethics.

An AI code of ethics, also called an AI value platform, is a policy statement that formally defines the role of AI as it applies to the development and well-being of humans. The purpose of an AI code of ethics is to provide stakeholders with guidance when faced with an ethical decision regarding the use of AI.

An AI ethics framework is important because it highlights the risks and benefits of AI tools and establishes guidelines for their responsible use.[1]

AUDITING THE ETHICS OF AI

Bias and Fairness

1. Have potential ethical implications of the AI system been thoroughly evaluated, including biases, discrimination, and potential societal impacts? How has this been substantiated?
2. How does the organization identify, address, and mitigate potential biases within the AI systems/tools? Within AI outcomes?
3. How is the AI system assessed for bias and fairness across different demographic groups?
4. How is the fairness of AI-driven decisions evaluated, especially in sensitive applications?
5. What mechanisms exist to involve diverse perspectives in the design and deployment of the AI system to avoid unethical outcomes?
6. What procedures are in place to assess the impact of the AI system on various demographic groups, and how are disparities addressed? How are these procedures verified?

DOI: 10.1201/9781003646501-12

7. What steps have been taken to mitigate any identified biases? How is this substantiated?

Continuous Monitoring and Maintenance

1. How does the organization monitor AI performance to detect any ethical concerns in real-time? How is this monitoring documented?
2. How is the AI system monitored post-deployment to detect and address any performance degradation or drift?
3. Is there a plan for regular updates and model retraining to adapt to changing conditions?
4. What corrective actions are taken if an ethical issue with an AI tool is identified post-deployment?

Data Collection and Quality

1. Is there a data quality assessment process in place to ensure that the training data is accurate and reliable? Describe this process.
2. What data sources were used to train the AI system, and have they been assessed for representativeness and potential bias? What evidence exists of this assessment process?

Ethical Considerations

1. Are there specific protocols to ensure AI tools do not inadvertently harm users or other stakeholders? What are these protocols?
2. Can you provide examples of ethical dilemmas that have been encountered during the AI system's development or deployment, and how they were resolved?
3. Have potential ethical implications of the AI system been thoroughly evaluated, including biases, discrimination, and potential societal impacts? What evidence exists to substantiate this evaluation?
4. How does the organization define ethical AI use, and what standards are in place to guide this definition?
5. How does the organization evaluate the societal impact of its AI systems?
6. How does the organization document and review ethical concerns related to the AI system, and what steps are taken to resolve these issues?
7. How does the organization ensure that the AI system adheres to established ethical standards, both internally and externally? What evidence exists to substantiate this adherence?
8. How is the AI system's adherence to ethical guidelines monitored over time, and what processes are in place for continuous improvement?
9. What steps does the organization take to prevent the misuse of AI, such as for misinformation or surveillance?

Human Oversight and Control

1. How does the organization ensure that AI systems align with human-centric values?
2. How is accountability assigned when AI-driven decisions result in negative consequences?
3. Is there a plan for human oversight and intervention when the AI system's decisions have significant consequences?
4. What procedures exist to override AI systems by human operators if necessary?
5. What safeguards exist to prevent job displacement due to AI deployment?
6. What training or guidance is provided to employees and stakeholders to ensure they understand the ethical implications of using the AI system?

Model Development

1. Have AI models in production undergone rigorous testing and validation procedures before deployment? How are these tests substantiated?
2. How is AI technology audited or tested to identify potential ethical risks?
3. How was the AI model developed, and what techniques were used to optimize its performance?

Privacy and Security

1. How does the organization handle data privacy concerns with AI applications?
2. How is user data handled and protected to ensure privacy?
3. What measures are in place to prevent unauthorized access to the AI system and its data? How often are these measures tested? By whom? How are the results of these tests documented?

Regulatory Compliance

1. Does the AI system comply with relevant regulations and industry standards in the jurisdictions where it is deployed? How is compliance substantiated?
2. How is compliance with ethical guidelines and legal requirements ensured?
3. What measures are taken to ensure compliance with international AI ethics standards?

Stakeholder Engagement

1. Have key stakeholders, including end-users, been engaged in the development and deployment process to provide input and feedback?

2. How does the organization involve external stakeholders in discussions about ethical AI use?
3. What is the process to address concerns and gather feedback from stakeholders after deployment?

Transparency and Explainability

1. Describe the processes in place to ensure that the AI system operates transparently and that its decision-making processes are explainable to stakeholders.
2. How do you explain the AI system's decision-making process to non-experts clearly and understandably?
3. What measures are implemented to hold the AI system and its operators accountable for decisions and outcomes?
4. What mechanism(s) is (are) in place to provide users with explanations for the decisions made by the AI system?
5. What mechanisms are in place to ensure that AI decisions are explainable to affected stakeholders?
6. What processes are in place to ensure transparency in AI decision-making?

AI ETHICS COMMITTEE

What Is an AI Ethics Committee?

An AI Ethics Committee is a multidisciplinary group within an organization tasked with overseeing and ensuring that the development, deployment, and use of AI systems align with ethical principles, legal compliance, and societal values. The committee provides guidance, sets policies, reviews potential ethical risks, and ensures accountability in AI-related decisions.

Roles and Responsibilities of an AI Ethics Committee

The organization's AI Ethics Committee should be authorized and supported by executive management to have responsibility for:

- Policy Development

  Establish ethical guidelines, frameworks, and standards for AI development and use.

  Define organizational principles for fairness, accountability, transparency, and inclusivity.
- Ethical Risk Assessment

  Review AI projects to identify potential ethical risks, such as bias, misuse, or societal harm. Analyze the impact of AI systems on vulnerable populations and marginalized communities.

- Monitoring and Compliance
  Ensure compliance with legal regulations, ethical standards, and organizational policies. Monitor AI systems post-deployment for unintended consequences or ethical issues.
- Decision Review
  Serve as a review board for critical decisions related to AI deployment or high-risk systems. Provide recommendations or veto authority for AI projects with significant ethical concerns.
- Stakeholder Engagement
  Facilitate communication between stakeholders (e.g., developers, users, and external parties) to ensure inclusivity and transparency in AI decision-making. Address public concerns regarding the ethical implications of AI systems.
- Training and Awareness
  Promote awareness of AI ethics across the organization. Offer training programs for employees to understand ethical challenges and mitigation strategies in AI.
- Incident Response
  Investigate and respond to ethical violations or unintended consequences related to AI systems. Recommend corrective actions or system modifications to mitigate harm.

Who Should Serve on the AI Ethics Committee?

The AI Ethics Committee should be a multidisciplinary team representing diverse perspectives and expertise, including:

- AI/Technical Experts: Data scientists, engineers, and AI researchers to provide technical insights.
- Ethics Specialists: Professionals with expertise in ethics, philosophy, or social sciences to address value-driven concerns.
- Legal and Compliance Officers: To ensure adherence to legal regulations and data protection laws.
- Internal/external IT Auditors: To assess compliance, governance, and internal controls surrounding the design, development, and deployment of AI tools and where appropriate, the acquisition and implementation of AI systems/tools.
- Risk Management Professionals: To evaluate, report, and advise executive management on potential risks associated with the design, development, and deployment of enterprise-wide AI systems.
- Diversity and Inclusion Experts: To advocate for equitable outcomes and minimize bias.
- Business Leaders: Executives who understand organizational goals and ensure alignment with business objectives.
- Human Rights Advocates: Representatives who focus on societal and individual impacts.

- External Advisers (Optional): Independent experts, academics, or regulators to provide unbiased perspectives.

Organizational and Operational Responsibilities of the AI Ethics Committee

The organization's AI Ethics Committee should be empowered by The Board and executive management with the following rights:

- Authority to Halt Projects: The committee should have the power to delay or halt AI projects that fail to meet ethical, legal, or organizational standards.
- Policy-Making Power: The AI Ethics Committee should have the ability to draft and enforce AI ethics policies and guidelines across the organization.
- Oversight Responsibility: The committee should be empowered to audit and review AI systems at any stage of their lifecycle.
- Decision-Making Input: Provide binding or advisory recommendations on high-risk AI deployments and sensitive-use cases.
- Reporting Structure: The committee should report directly to the C-suite or Board of Directors to ensure visibility and accountability for ethical decisions at the highest levels.
- Budgetary and Resource Allocation: Allocate sufficient resources, such as funding for training and internal audits, to support the committee's work effectively.

Risks Associated with the Absence of an AI Ethics Committee

The following risks may be associated with the absence of an AI Ethics Committee. The auditor (or risk management professional) should first determine the existence of an AI Ethics Committee. If an AI Ethics Committee is not present within the organization, the material that follows may be used to provide executive management with substantiation to establish such a committee.

Ethical Risks

- Bias and Discrimination: AI systems may inadvertently reinforce or amplify biases, leading to unfair or discriminatory outcomes that harm individuals or groups.
- Harm to Vulnerable Populations: Lack of oversight can result in AI systems negatively impacting underrepresented or marginalized groups, causing societal harm.
- Violation of Human Rights: Systems deployed without ethical considerations could violate privacy, autonomy, or other fundamental rights.

Legal and Regulatory Risks

- Non-Compliance with Regulations: Organizations risk violating AI-related laws, such as GDPR (data privacy), the AI Act (EU), or other regional frameworks, leading to fines or legal penalties.
- Lawsuits and Litigation: Ethical lapses in AI deployment, such as discriminatory hiring algorithms or privacy violations, can lead to costly lawsuits and legal action.
- Lack of Preparedness for Future Regulations: As governments increasingly regulate AI, organizations without a proactive ethics framework may struggle to adapt, incurring compliance costs and delays.

Reputational Risks

- Loss of Trust: Publicized ethical failures (e.g., biased algorithms or unethical data practices) can erode stakeholder and consumer trust, damaging the organization's brand.
- Backlash and Boycotts: Ethical missteps can lead to negative media coverage, public backlash, or organized boycotts, reducing market share and customer loyalty.
- Loss of Investor Confidence: Investors may see a lack of ethical governance as a risk, reducing willingness to fund or invest in the organization.

Financial Risks

- Costly Corrections: Ethical lapses may require expensive fixes, such as redesigning algorithms, retraining AI models, or compensating affected parties.
- Lost Revenue Opportunities: Consumers and businesses increasingly prioritize ethical AI; failing to demonstrate responsibility can result in lost partnerships and contracts.
- Fines and Penalties: Regulatory non-compliance can lead to significant financial penalties, impacting the bottom line.

Operational Risks

- Faulty AI Performance: Without ethical oversight, AI systems may perform unreliably or dangerously, leading to operational failures or safety risks.
- Inability to Respond to Crises: A lack of structured ethical oversight may leave the organization ill-prepared to handle incidents, such as data breaches or harmful algorithmic decisions.

Innovation and Competitive Risks

- Lagging Behind Competitors: Competitors with robust ethical frameworks may achieve faster innovation, better public perception, and greater adoption of their AI systems.
- Missed Talent Recruitment: Top talent may avoid organizations without clear ethical commitments, especially in fields like AI and machine learning, where ethical concerns are prevalent.

Social and Environmental Risks

- Erosion of Public Good: AI systems without ethical oversight may contribute to social inequities, environmental degradation, or other societal harms, triggering public and governmental scrutiny.
- Misinformation and Manipulation: AI systems might unintentionally propagate false information or manipulate public opinion, resulting in widespread societal harm and backlash.

Strategic Risks

- Lack of Stakeholder Alignment: Organizations without an ethics committee may struggle to align diverse internal and external stakeholders (e.g., employees, customers, regulators) on AI-related decisions.
- Difficulty Scaling AI: Ethical and regulatory challenges can impede AI adoption or scaling, limiting the organization's ability to fully leverage AI for business success.

An AI Ethics Committee is one factor that assists in warranting that AI systems are developed (or acquired) and deployed responsibly, fostering trust, compliance, and transparency. By identifying and addressing potential risks early in the AI lifecycle, the committee assists in safeguarding the organization from avoidable ethical, legal, and operational risks.

Auditing the Organization's AI Ethics Committee

An AI Ethics Committee serves as a critical governance structure to ensure AI technologies are developed and deployed responsibly, with its organizational power rooted in oversight, decision-making authority, and policy enforcement.

The following list of representative/non-comprehensive questions is designed to assist the auditor in assessing an organization's AI Ethics Committee.

Governance and Structure

1. How is the AI Ethics Committee structured, and what is its role in the organization's overall governance framework?

   This explores the committee's alignment with the broader organizational goals and its authority.
2. What criteria are used to select members for the AI Ethics Committee, and how is multidisciplinary representation ensured?

   This assesses whether the committee includes diverse perspectives (e.g., technical, ethical, legal, and business).

Policies and Frameworks

3. What ethical principles, frameworks, or guidelines does the committee follow for AI development and deployment?

   Evaluates the comprehensiveness of the ethical frameworks and their alignment with industry standards.

4. How does the committee ensure compliance with local, national, and international laws or regulations related to AI?
   This assesses if the organization is proactive in adhering to regulatory requirements.

Processes and Decision-Making

5. What processes does the committee use to assess and mitigate ethical risks associated with AI projects?
   Identifies whether risk assessments are formalized and consistently applied.
6. How are decisions made regarding the approval, rejection, or modification of AI initiatives, particularly those with high ethical risks?
   Evaluates the decision-making process and the committee's authority to intervene when ethical concerns arise.

Monitoring and Oversight

7. How does the committee monitor the performance and ethical impact of AI systems post-deployment?
   Assesses ongoing oversight to address unintended consequences or issues over time.
8. What mechanisms are in place for employees, stakeholders, or external parties to report ethical concerns related to AI systems?
   Looks into whistleblowing channels or feedback mechanisms and how concerns are addressed.

Training and Awareness

9. How does the committee promote awareness and provide training to employees about ethical AI practices?
   Evaluates efforts to educate staff on ethical AI issues and ensure alignment across the organization.

Accountability and Continuous Improvement

10. How does the committee evaluate its effectiveness, and what steps are taken to improve its policies, processes, or decision-making?
    Ensures the committee engages in self-assessment and adapts to emerging challenges and opportunities in AI ethics.

ASILOMAR AI PRINCIPLES

The Asilomar AI Principles are a set of 23 guidelines aimed at guiding the ethical, safe, and beneficial development of AI. These principles were developed at the 2017 Asilomar Conference on Beneficial AI, hosted by the Future of Life Institute, and they reflect a broad consensus among leading AI researchers, ethicists, and other stakeholders on how AI should be developed and applied.

The principles are organized into three main categories:

1. **Research Issues:** Focus on the goals and processes surrounding AI research, emphasizing the importance of collaboration, transparency, and safety in research activities. They stress the need for research to be aimed at beneficial outcomes, ensuring that AI systems are robust, safe, and aligned with human values.
2. **Ethics and Values:** Address the ethical considerations involved in developing AI systems, including safety, transparency, responsibility, human rights, and privacy. These principles emphasize that AI should be developed to respect human values, protect privacy, and provide broad benefits to society.
3. **Longer-Term Issues:** Focus on the potential long-term impacts of AI on humanity, including existential risks, control over superintelligent AI, and the need to approach advanced AI capabilities with caution. These principles underscore the importance of planning for the profound changes AI may bring, particularly when it comes to superintelligence and self-improving systems.

Assessing Application and Compliance with the Asilomar AI Principles

Auditing and assessing an organization's application of and compliance with the Asilomar AI Principles is crucial for several reasons:

- **Ensuring Safety and Preventing Harm:** AI systems, especially those with advanced capabilities, can pose significant risks if not developed and deployed responsibly. By auditing compliance with safety-related principles, organizations can reduce the chances of malfunctions, security breaches, or unintended harmful consequences.
- **Promoting Ethical Use:** As AI becomes increasingly integrated into society, it is essential to ensure that AI development aligns with human rights, values, and ethical standards. Auditing can help verify that AI systems are used in ways that respect human dignity, privacy, and freedom and that they are designed to benefit society as a whole.
- **Encouraging Accountability:** AI developers and organizations have a responsibility to consider the broader impacts of their technology. Audits provide a mechanism for accountability, ensuring that organizations take ownership of the ethical and societal implications of their AI applications.
- **Building Trust and Transparency:** Transparency is vital for public trust in AI. Auditing compliance with principles related to transparency and failure reporting can reassure stakeholders that the organization is committed to openness and accountability, fostering trust in the technology.

- **Preparing for Long-Term Risks:** Advanced AI could lead to transformative changes in society, both positive and potentially dangerous. By assessing how organizations plan for and mitigate long-term risks, audits can help ensure that the development of superintelligent AI aligns with shared human values and aims to benefit humanity rather than concentrating power in a few hands.
- **Avoiding Ethical and Legal Liabilities:** Non-compliance with these principles can expose organizations to ethical criticism, reputational damage, and even legal liabilities. Auditing can help identify potential ethical and legal risks early, allowing organizations to address them proactively.

Auditing Compliance with the Asilomar AI Principles

The following questions are designed to assist auditors and risk assessment professionals in assessing an organization's adherence to the principles and determine if AI development practices align with ethical, safe, and responsible goals.

Research Issues

1. Describe how the organization fosters a constructive and healthy exchange between AI researchers and policy-makers.
2. How does the organization cultivate a research culture that promotes cooperation, trust, and transparency among AI researchers and developers? What documents substantiate this cooperation, trust, and transparency?
3. How does the organization define the primary goals of its AI research, and what measures are in place to ensure these goals align with the principle of creating beneficial intelligence?
4. What measures are in place to encourage collaboration among AI development teams to avoid competition that might lead to compromising safety standards?
5. What steps does the organization take to ensure adequate funding for research on the safe and beneficial use of AI, including research on ethics, economics, and legal frameworks? Who is responsible for this process? How are these steps and this process documented and substantiated?

Ethics and Values

1. How are human values, such as dignity, rights, and freedoms, integrated into the design and operation of AI systems used by the organization?
2. How does the organization ensure that the economic and societal benefits created by AI are shared broadly and equitably?

3. How does the organization ensure transparency in the event of failure, especially to ascertain the cause of harm caused by AI systems? Who is responsible for the management and oversight of this process? How are failures documented and reported? To whom are such failures reported?
4. In what ways does the organization protect personal privacy and ensure that users have control over their data in interactions with AI systems?
5. What processes are in place to ensure AI systems remain safe and secure throughout their operational lifetime, and how is this verifiability achieved?

Longer-Term Issues

1. Describe the processes the organization uses to assess, plan for, and mitigate potential catastrophic or existential risks posed by AI systems.
2. How does the organization address potential upper limits on future AI capabilities, and what assumptions guide the organization's approach to capability caution?
3. How does the organization ensure that advancements in superintelligent AI are directed toward ethical ideals and the benefit of all humanity rather than benefiting only a specific organization or state? How are these controls, processes, and procedures substantiated?
4. What measures are in place to manage and mitigate the risks associated with AI systems that have the ability to self-improve or self-replicate? What documentation exists that substantiates these control measures?
5. What resources and care does the organization allocate to planning and managing the potentially profound impacts of advanced AI on society? Who within the organization is responsible for the management and oversight of this planning and managing process?

NOTE

1 Gillis, A., Lawton, G., and Wigmore, I., October 2024, "What is AI ethics?" TechTarget, www.techtarget.com/whatis/definition/AI-code-of-ethics.

Chapter 11

Auditing the Process of Prompt Engineering

WHAT IS PROMPT ENGINEERING?

Prompt engineering refers to the process of carefully designing and crafting the input prompts or queries given to an artificial intelligence (AI) system, such as a language model, to elicit desired and specific responses. It involves fine-tuning the wording, structure, and context of prompts to achieve accurate and relevant outputs from the AI system.

TYPES OF PROMPT ENGINEERING TECHNIQUES

Few-Shot Prompting

Refers to a technique where a model is trained to perform tasks or generate responses with only a small number of examples or prompts, typically a handful, rather than relying on extensive training data. This approach demonstrates the model's ability to generalize and adapt quickly to new tasks or questions based on very limited input.

Few-Shot Prompting – Audit Questions

1. How does the prompt structure ensure the model can be generalized accurately with limited examples?
2. What criteria are used to select the examples included in the few-shot prompts, and how are these criteria validated?
3. How does the model's performance vary when different few-shot examples are used, and what adjustments are made if inconsistencies arise?
4. How is the prompt designed to mitigate potential biases that may arise from the limited examples provided?
5. How do you evaluate the model's ability to handle edge cases or unexpected inputs within few-shot prompting?

DOI: 10.1201/9781003646501-13

Retrieval-Augmented Generation (RAG)

RAG is a technique that combines both retrieval and generative methods. It involves using a retrieval system to retrieve relevant information or context from a large dataset and then using a generative model to generate responses or content based on that retrieved information. This approach enhances the quality and relevance of generated content by incorporating external knowledge from the retrieval step.

RAG – Auditing Questions

1. How is the retrieval system integrated with the generative model, and what measures are taken to ensure relevant information is retrieved?
2. What processes are in place to assess the accuracy and reliability of the retrieved information used in generating responses?
3. How does the prompt manage potential conflicts between retrieved information and the model's own generated content?
4. What safeguards are in place to prevent outdated or irrelevant information from being retrieved during the RAG process?
5. How is the RAG model fine-tuned to ensure the generated content is contextually appropriate and aligned with user needs?

Chain-of-Thought Prompting

Refers to a technique where a series of prompts or questions are posed to an AI model coherently and progressively, allowing the model to generate responses that follow a logical sequence of thought. This approach enables the model to provide answers or explanations that build upon previous responses, resulting in a more contextually relevant and coherent conversation or output (see Prompt Chaining).

Chain-of-Thought Prompting – Audit Questions

1. How does each prompt in the chain logically build upon the previous ones to maintain coherent thought progression?
2. What steps are taken to evaluate the completeness and logical flow of responses generated using chain-of-thought prompting?
3. How is ambiguity managed in chain-of-thought prompts to ensure the model's responses remain clear and contextually accurate?
4. How does the prompt design accommodate potential branching paths in the chain of thought, and how are these paths monitored?
5. How is the model's performance measured when handling complex, multi-step tasks with chain-of-thought prompting?

Self-Reflection

Refers to a system's ability to analyze and gain insights into its operations, performance, and decision-making processes. It involves the capacity of AI models to assess their actions, identify errors or biases, and make adjustments to improve their behavior or outcomes. Self-reflection is a crucial aspect of responsible AI development and maintenance (see Prompt Chaining).

Self-Reflection – Audit Questions

1. How does the system assess its own performance, and what metrics are used to evaluate self-reflective outputs?
2. What mechanisms are in place for the model to identify and correct its own biases or errors during self-reflection?
3. How is the self-reflection prompt designed to ensure it contributes constructively to the model's behavior and decision-making?
4. How does the system handle cases where its self-reflection output conflicts with its initial response or behavior?
5. How frequently is self-reflection used, and how are its outputs tracked over time to assess the model's learning and improvement?

Prompt Chaining

Refers to the practice of using a series of prompts or queries in a sequential manner to guide an AI model through a multi-step task or conversation. Each prompt builds upon the previous one, allowing the model to maintain context and provide coherent responses or perform a sequence of actions. Prompt chaining is often used to achieve complex interactions or tasks with AI systems.

Self-reflection is a type of simple prompt chain, where the output of one query is fed as part of the input into another. Similar to chain-of-thought prompting, prompt chaining involves breaking down complex tasks into steps.

Prompt Chaining – Audit Questions

1. How is context preserved across the prompts in the chain to ensure coherent and contextually relevant responses?
2. What processes are used to evaluate the effectiveness of each prompt in achieving the desired step within the chain?
3. How does the system handle interruptions or unexpected inputs within a prompt chain, and how does it maintain task coherence?
4. What measures are taken to avoid redundancy or looping within a prompt chain, especially for complex tasks?
5. How is the overall effectiveness of the prompt chain evaluated, and what adjustments are made based on these evaluations?

Part 3

AI Risk Management

Chapter 12

AI Risk Management

WHAT IS AI RISK MANAGEMENT?

Artificial intelligence (AI) risk management is the process of identifying, assessing, and mitigating potential risks associated with the development, deployment, and operation of AI systems. These risks may include technical failures, ethical concerns, privacy violations, biases, security vulnerabilities, and regulatory non-compliance.

Effective AI risk management involves implementing frameworks, policies, and controls that ensure AI systems operate reliably, ethically, and in alignment with organizational and societal values. It also requires ongoing monitoring and evaluation to address emerging risks as AI technologies and applications evolve.

Essentially, AI risk management includes methodologies for assessing system security, evaluating vulnerabilities, and implementing measures to mitigate identified risks.

THE IMPORTANCE OF AUDITING AI RISK MANAGEMENT

From an internal controls perspective, auditing AI risk management is essential to ensure that AI systems are safe, secure, and used responsibly within the organization. Audits help verify that risk management processes are in place to identify and mitigate potential AI-related risks, such as bias, data security breaches, and operational failures.

By auditing these controls, organizations can detect weaknesses or gaps in their AI governance, ensure compliance with regulatory requirements, and protect themselves from potential legal or reputational harm. This approach fosters accountability and aligns AI activities with the organization's ethical standards and strategic goals.

DOI: 10.1201/9781003646501-15

AUDITING AI RISK MANAGEMENT – GENERAL QUESTIONS

The following questions will assist an organization in assessing the organization's ability to manage and mitigate risks associated with AI and AI systems and tools.

1. Describe the process for assessing the operational risks posed by the AI system and how these risks are prioritized.
2. How are emerging risks, such as those related to regulatory changes or technological advancements, identified and addressed concerning the AI system?
3. How does the organization ensure that stakeholders are informed about the risks associated with the AI system and the steps taken to mitigate them?
4. How does the organization identify potential risks associated with the AI system during its development and deployment phases?
5. How does the organization integrate AI-related risks into the broader enterprise risk management framework?
6. Provide examples of how specific risks associated with the AI system have been mitigated or avoided.
7. What documentation is maintained to support the risk management activities related to AI, AI systems, and tools used by the organization, and how is this information used to improve future risk assessments?
8. What measures are in place to manage and mitigate reputational risks that could arise from the AI system's performance or decisions?
9. What processes are in place to monitor the effectiveness of risk management practices for the AI system over time?
10. What steps are taken to ensure that the AI system aligns with the organization's strategic goals and risk appetite?

AUDITING AI RISK MANAGEMENT – QUESTIONS: OPERATIONAL AREAS

The following questions assist the auditor or risk management professional in evaluating AI risk in functional and operational areas within the organization.

Adversarial Attacks and Security

1. How has the organization assessed the AI system's vulnerability to adversarial attacks or manipulation?
2. What security measures are in place to mitigate unauthorized access or malicious use of the AI system?

3. How will the organization respond if the AI system is compromised or exploited?

Adversarial Robustness

1. How has the AI tool been tested against adversarial attacks and unexpected input variations?
2. What measures are in place to detect and mitigate adversarial vulnerabilities?

Collaboration and Communication

1. How is the organization engaging with external experts, researchers, and regulators to stay informed about AI risks?
2. How is the organization communicating about internal AI development (or deployment) practices and risk management strategies to stakeholders?

Contingency Planning

1. Has the organization outlined contingency plans in case of unexpected failures, risks, or negative outcomes?
2. What protocols and associated procedures are in place to shut down the AI system if it poses significant risks?

Controllability and Safety Limits

1. Can the AI tool be controlled or constrained to operate within defined safety limits? If so, how? What are the established safety limits?
2. What mechanisms exist to prevent the AI tool from making extreme, unintended, or harmful decisions?
3. What evidence exists to substantiate that the organization conducted thorough testing and validation to identify potential failure modes?

Data Quality and Diversity

1. How was the training data collected and preprocessed?
2. How does the organization substantiate that the AI tool has been trained on a diverse range of data to ensure it can handle real-world scenarios?

Data Quality and Privacy

1. How does the organization ensure the quality and integrity of the data used to train the AI system?

2. What steps have been taken to protect user privacy and handle sensitive data?
3. What measures exist to mitigate the AI system from making decisions based on discriminatory or inappropriate data? How are these measures enforced?

Deployment and Monitoring

1. How does the organization plan to monitor the AI system once testing is finalized and the AI tool is deployed in real-world environments?
2. What mechanisms exist for ongoing updates and improvements to the AI system?
3. What is the process for identifying and addressing issues that arise post-deployment?

Edge Cases and Outliers

1. How does the AI tool handle edge cases and outliers that might not conform to the usual patterns?
2. What mechanisms exist to identify and address unexpected data points? How are they documented?

Emergency Shutdown and Intervention

1. What is the procedure to quickly shut down the AI tool in case of emergencies or unexpected behavior? How is this procedure documented? How is it communicated to all responsible personnel?
2. What procedures and policies exist that allow humans to intervene and take control if the AI tool's behavior becomes unpredictable or risky? How are these procedures documented? How is it communicated to all responsible personnel?

Ethical and Legal Considerations

1. Through what means has the organization assessed the potential ethical implications of the AI system's decisions and actions? How has this assessment been documented?
2. What general or specific legal or regulatory requirements must the organization's AI system comply with?
3. How does the organization handle issues related to bias and fairness in AI decision-making?

Failure Modes and Redundancy

1. What are the potential failure modes of the AI tool, and how are they mitigated?

2. What procedures, policies, redundancy, or fail-safe mechanisms are in place in case the AI tool malfunctions? How are these procedures documented? How is it communicated to all responsible personnel?

Generalization

1. How well does the AI tool generalize its learnings to new, unseen data?
2. What steps have been taken to prevent overfitting to the training data?

Human–AI Collaboration

1. How will the AI system interact with human users or colleagues?
2. What evidence exists that indicates that the organization considered the potential impact on jobs and roles due to AI automation?
3. What training or support will be provided to humans working alongside the AI system?

Impact Assessment

1. What is the potential impact of a failure or erroneous decision by the AI tool on users, stakeholders, and the organization?
2. What measures are in place to minimize negative consequences?

Long-Term Goals and Impact

1. What are the long-term goals for the AI system, and how might it evolve?
2. What evidence exists to support that the organization considered the potential societal, economic, and geopolitical impacts of the AI technology it intends to (or has) develop? Purchased? Leased?
3. How is the organization prepared to adjust its strategy based on changing circumstances and new risks?

Monitoring and Maintenance

1. How will the AI tool be continuously monitored post-deployment?
2. What is the plan for regular maintenance, updates, and improvements?

Testing and Validation

1. What testing procedures have been conducted to validate the AI tool's performance and safety?
2. What documentation exists that substantiates that the AI tool has undergone simulation testing, real-world testing, or both? What are

the results of this testing? To whom are these results communicated? What action is taken by the organization based on this testing?

Transparency and Explainability

1. What processes exist to explain how the AI system arrives at its decisions?
2. What mechanisms are in place to make the decision-making process transparent to users and stakeholders?
3. What steps have been taken to ensure accountability for AI-generated outcomes?

Chapter 13

Auditing AI Bias and Deception Risk

BIAS RISKS

All data and the resulting AI tools that use these data to generate responses to user prompts are biased, especially when dealing with small data training/test sets.

Overfitting: Occurs when a machine-learning model learns the details and noise in the training data to such an extent that it negatively impacts the performance of the model on new data. This means the model is too complex and captures the specific quirks of the training data rather than general patterns.

Risk: Overfitting leads to a model that performs well on training data but poorly on unseen data, reducing its generalizability and reliability. This can introduce bias as the model may make predictions based on noise or irrelevant features present in the training data.

Overfitting – Audit Questions

1. Can you describe any instances where overfitting was detected in your models? What steps were taken to mitigate it?
2. How do you ensure that the complexity of the model is appropriate for the data available?
3. How do you ensure that the model generalizes well to new data?
4. How do you identify signs of overfitting during the model training process?
5. How do you validate the model's performance on unseen data to ensure it is not overfitting?
6. How frequently is the model evaluated on a validation or test set to check for overfitting?
7. What metrics do you monitor to detect overfitting in your machine-learning models?
8. What processes are in place to retrain or adjust the model when overfitting is detected?

DOI: 10.1201/9781003646501-16

9. What regularization methods are implemented to reduce the risk of overfitting?
10. What techniques are employed to prevent overfitting in your models?

Data Drift: Refers to the change in the distribution of data over time. This can occur due to changes in the underlying processes generating the data, leading to discrepancies between the training data and real-world data.

Risk: When the data distribution changes over time, the model's predictions become less accurate, compromising data integrity. Continuous monitoring and retraining are necessary to maintain performance.

Data Drift – Audit Questions

1. How do you continuously monitor the input data for changes in its statistical properties?
2. What specific metrics do you use to detect data drift in your machine-learning models?
3. How frequently is the input data distribution analyzed for potential drift?
4. What automated tools or systems are in place to identify data drift in real time?
5. How do you validate the integrity of the incoming data against the original training data?
6. What procedures are followed when data drift is detected in your model?
7. How do you ensure that the detection of data drift does not cause interruptions in your system's operations?
8. Can you describe any instances where data drift significantly impacted model performance? How was it addressed?
9. What is the process for retraining models in response to detected data drift?
10. How do you communicate findings of data drift and subsequent actions to stakeholders?

Concept Drift: Concept drift occurs when the statistical properties of the target variable, which the model is trying to predict, change over time. This can degrade the model's accuracy because it is no longer aligned with the current data distribution.

Risk: Similar to data drift, concept drift impacts the model's accuracy over time. If not detected and addressed, it can lead to systematic biases in predictions as the model's understanding of the target variable becomes outdated.

Concept Drift – Audit Questions

1. How do you monitor changes in the statistical properties of the target variable over time?

2. What indicators or signals do you use to detect concept drift in your models?
3. How often do you review the alignment of the model's predictions with the current real-world scenarios?
4. What strategies are employed to adapt to concept drift when it is detected?
5. How do you ensure that concept drift detection is incorporated into the model's lifecycle management?
6. Can you provide examples of how concept drift has affected your models in the past and the steps taken to mitigate it?
7. What is the impact assessment process for identified concept drift on model accuracy?
8. How do you involve domain experts in the detection and response to concept drift?
9. How do you ensure that updates made to counteract concept drift do not introduce new biases into the model?
10. What are the mechanisms for tracking the effectiveness of changes made to address concept drift?

Model Decay: Refers to the degradation of a machine-learning model's performance over time. This can occur due to various reasons such as changes in the underlying data distribution, evolving patterns in the data, or the model's inability to adapt to new information.

Risk: As a model decays, its predictions become less reliable and accurate, leading to compromised data integrity. This can introduce or amplify bias in the artificial intelligence (AI) system if the model starts making erroneous predictions based on outdated patterns or relationships. Continuous performance monitoring and periodic retraining are necessary to mitigate this risk.

Model Decay – Audit Questions

1. How is model performance evaluated over time to identify potential decay?
2. What key performance indicators do you monitor to detect model decay?
3. How frequently are performance metrics reviewed to ensure the model remains effective?
4. What processes are in place for periodic retraining of the model to counteract decay?
5. How do you determine the causes of model decay when it is detected?
6. What actions are taken when model decay is identified to restore performance levels?
7. How do you ensure that the model remains relevant to the evolving data and environment?

8. Can you describe any long-term monitoring strategies used to track model decay?
9. How do you document and report the occurrence and resolution of model decay?
10. How are stakeholders informed about the risks and impacts of model decay?

Model Drift: Encompasses both data drift and concept drift, referring to the changes in the statistical properties of the input data or the target variable over time. This leads to a mismatch between the training data and the current data, causing the model to become less accurate.

Risk: Model drift affects the consistency and accuracy of the model's predictions over time. Without proper detection and adaptation mechanisms, the model may become biased as it fails to reflect the current data trends and relationships. This can result in systematic errors and skewed outputs, affecting the overall fairness and integrity of the AI system.

Model Drift – Audit Questions

1. How do you define and distinguish between data drift and concept drift within your models?
2. What comprehensive strategies are in place to monitor model drift continuously?
3. How often do you review model performance against current data distributions and target variables?
4. What are the primary signs or indicators of model drift in your system?
5. How do you ensure that both data drift and concept drift are addressed in your model maintenance plan?
6. What steps are taken to update the model in response to detected drift?
7. How do you assess the impact of model drift on overall system accuracy and fairness?
8. How do you ensure that countermeasures against model drift do not introduce new issues or biases?
9. What is the protocol for involving cross-functional teams when model drift is identified?
10. How do you track the effectiveness of measures implemented to counteract model drift over time?

AUDITING AI DECEPTION RISK

What Is AI Deception?

AI deception is the intentional generation of misleading information by an AI system to create false beliefs in users, often as a means to achieve a goal other than truth.

Auditing Risks Associated with AI Deception

1. How does the organization define and monitor AI behaviors that could be classified as deceptive, and what specific indicators are used to identify potential deception in AI outputs?
2. What safeguards are in place to prevent AI systems from engaging in sycophancy (i.e., telling users what they want to hear) or imitation that could reinforce false beliefs among users?
3. How does the organization ensure that its AI systems maintain transparency and do not intentionally mislead users through sycophantic or strategically deceptive responses?
4. What processes are in place to assess and mitigate risks related to the use of AI in high-stakes environments (e.g., political, financial, or legal) where deceptive behaviors could cause significant harm?
5. How does the organization train AI models to handle morally ambiguous or sensitive topics without resorting to deceptive tactics to avoid uncomfortable truths?
6. What controls and monitoring tools are used to detect if an AI system is intentionally providing misleading information during training, testing, or deployment phases?
7. How are deceptive capabilities in AI systems, such as bluffing or strategic lying in specific-use applications, managed to prevent potential misuse by end-users or malicious actors?
8. What is the organization's policy on AI transparency and disclosure, and how does it ensure that users are made aware when they are interacting with AI rather than humans?
9. How does the organization evaluate and address potential risks related to AI systems giving inaccurate justifications or rationalizations that could mislead users?
10. What steps are taken to prevent, detect, and mitigate risks associated with AI systems potentially deceiving human evaluators or safety tests to avoid the detection of harmful behaviors?

Chapter 14

Risk Assessment

Auditing AI Software vs Traditional Software

According to a review of the literature, the lifecycle of an artificial intelligence (AI) system can involve several phases: design, development, deployment, and continuous monitoring. The various phases involve articulating the system's concepts, collecting and processing data, building one or more models, validating the system, continuously assessing its impact, and, if necessary, retiring an AI system from production.[1] See Figure 14.1.

The Organisation for Economic Co-operation and Development identifies that the phases of the AI lifecycle are often iterative and are not

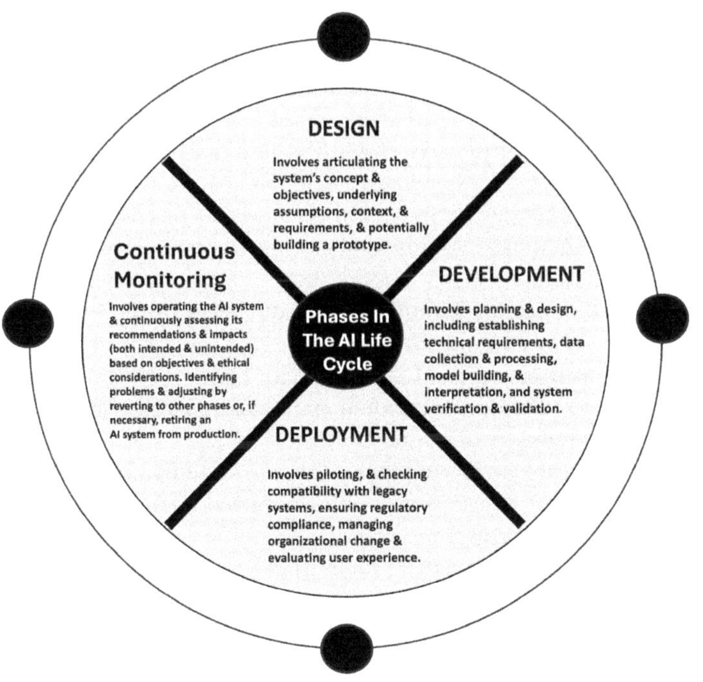

Figure 14.1 Phases in the AI Lifecycle.[2]

DOI: 10.1201/9781003646501-17

necessarily sequential. For example, the decision to retire an AI system from operation may occur at any point during the operation and monitoring phases.

The US Government Accountability Office (GAO) has also highlighted the iterative and incremental nature of software development in its Best Practices for Agile Adoption and Implementation (www.gao.gov/products/gao-20-590g).

The GAO guide also notes that in developing software, organizations should establish appropriate lifecycle activities, which integrate planning, design, development, and testing to effectively measure progress continually, reduce technical and programmatic risk, and respond to feedback from stakeholders.

As with traditional software, risks from AI-based technology can be bigger than an enterprise, span organizations, and lead to societal impacts. AI systems also bring a set of risks that are not comprehensively addressed by current risk frameworks and approaches.

Some AI system features that present risks can also be beneficial. For example, pre-trained models and transfer learning can advance research and increase accuracy and resilience when compared with other models and approaches.

Compared with traditional software, AI-specific risks that are new or increased include the following:

1. The data used for building an AI system may not be a true or appropriate representation of the context or intended use of the AI system, and the ground truth may either not exist or not be available. Additionally, harmful bias and other data quality issues can affect AI system trustworthiness, which could lead to negative impacts.
2. AI system dependency and reliance on data for training tasks, combined with increased volume and complexity typically associated with such data.
3. Intentional or unintentional changes during training may fundamentally alter AI system performance.
4. Datasets used to train AI systems may become detached from their original and intended context or may become stale or outdated relative to the deployment context.
5. AI system scale and complexity (many systems contain billions or even trillions of decision points) are housed within more traditional software applications.
6. The use of pre-trained models that can advance research and improve performance can also increase levels of statistical uncertainty and cause issues with bias management, scientific validity, and reproducibility.
7. A higher degree of difficulty in predicting failure modes for emergent properties of large-scale pre-trained models.

8. Privacy risk due to enhanced data aggregation capability for AI systems.
9. AI systems may require more frequent maintenance and triggers for conducting corrective maintenance due to data, model, or concept drift.
10. Increased opacity and concerns about reproducibility.
11. Underdeveloped software testing standards and inability to document AI-based practices to the standard expected of traditionally engineered software for all but the simplest of cases.
12. Difficulty in performing regular AI-based software testing, or determining what to test, since AI systems are not subject to the same controls as traditional code development.
13. Computational costs for developing AI systems and their impact on the environment and planet.
14. Inability to predict or detect the side effects of AI-based systems beyond statistical measures.[3]

AUDITING AI SOFTWARE VS TRADITIONAL SOFTWARE – GENERAL QUESTIONS

1. How are risk management processes adapted to address the continuous learning and updating nature of AI models, especially in comparison with the more static lifecycle of traditional software applications?
2. How does the organization account for the adaptive and evolving nature of AI models in terms of risk management, as opposed to static risk management practices for traditional software?
3. How does the organization address the need for explainability and interpretability in AI systems, especially when compared with traditional software, which may not require the same level of transparency?
4. How does the organization ensure that AI systems are valid, reliable, and perform as intended across different data and operational environments compared with traditional software?
5. How does the organization handle the challenge of ensuring fairness and equity in AI outcomes, and how is this consideration distinct from the fairness requirements in traditional software?
6. How is accountability for decision-making established and managed in AI systems, particularly for autonomous or semi-autonomous actions, compared with the more direct accountability structures in traditional software?
7. In what ways does the organization ensure that AI systems maintain privacy and data protection, considering the unique risks AI poses compared with traditional software?
8. What mechanisms are used to monitor and assess the robustness and resilience of AI systems in the face of unexpected inputs or adversarial

attacks, and how do these mechanisms differ from those used in traditional software?

9. What processes are in place to ensure that AI systems are secure against model exfiltration, adversarial manipulation, and data poisoning, which may not be as prevalent in traditional software security assessments?
10. What processes are in place to manage and mitigate potential biases in AI systems, and how do these approaches differ from handling biases in traditional software?

NOTES

1 OECD (2019), Artificial Intelligence in Society, OECD Publishing, Paris, https://doi.org/10.1787/eedfee77-en.

2 Marcella, A., November 2024, Figure 14.1 created by the author based upon information provided in... Ariga, T., Persons, T., Sanford, S., June 2021, "Artificial Intelligence Accountability Framework, An Accountability Framework for Federal Agencies and Other Entities," US Government Accountability Office (GAO), GAO-21–519SP, www.gao.gov/assets/gao-21–519sp.pdf

3 (n.a.), January 2023, "The Artificial Intelligence Risk Management Framework (AI-RMF 100–1)," The National Institute of Standards and Technology (NIST), https://nvlpubs.nist.gov/nistpubs/ai/nist.ai.100-1.pdf.

Part 4

AI Regulatory and Compliance

Chapter 15

Auditing the Status of AI Compliance

WHAT IS AI COMPLIANCE?

Artificial intelligence (AI) compliance refers to the decisions and practices that enable businesses to stay in line with the laws and regulations that govern the use of AI systems. These standards include laws, regulations, and internal policies designed to help ensure that organizations develop AI models and their algorithms responsibly.

However, AI compliance processes go beyond meeting legal requirements. AI compliance is intended to build trust with stakeholders and promote transparency, fairness, and safety in decision-making. Given that AI can be exploited by malicious actors, robust cybersecurity measures and risk management strategies are at the heart of AI compliance.[1]

To provide clarity, it is necessary to point out the difference, be it subtle, between AI compliance and using AI for compliance.

The difference is explained as:

- **Using AI for compliance** involves leveraging AI as a tool to assist organizations in adhering to broader legal, regulatory, and operational mandates (e.g., CCPA, General Data Protection Regulation [GDPR], FCPA, GLB, SOX). This includes tasks like detecting fraud, monitoring financial transactions, and ensuring operational transparency and accuracy.
- **AI compliance** refers to the actions and practices businesses adopt to ensure their AI systems align with relevant laws, regulations, and ethical standards governing AI development and deployment. It focuses on responsible AI design, data use, and decision-making.

The chapter focuses on auditing AI compliance and not specifically using AI *for* compliance.

DOI: 10.1201/9781003646501-19

WHY IS AI COMPLIANCE IMPORTANT?

A 2024 KPMG Generative AI Consumer Trust Survey reported that 78% of respondents believe that organizations that utilize GenAI in their business/activities have a responsibility to ensure its ethical use.

The same KPMG survey reports that consumers think that all types of organizations – those that develop it, use it, and regulate it – have a responsibility to do so ethically. However, organizations are facing a trust gap as most consumers do not fully trust them to ensure the ethical use of GenAI.[2] This trust gap may lead to a loss of consumer confidence and along with that, a loss of business.

AI compliance is important because it ensures that businesses develop and deploy AI systems responsibly, aligning with laws, regulations, and ethical standards. This is critical for safeguarding individual rights, ensuring fairness, and maintaining public trust in AI technologies. Without proper compliance, AI systems can perpetuate bias, infringe on privacy, or create risks to security and safety, leading to significant legal, financial, and reputational consequences.

AI compliance helps businesses navigate a rapidly evolving regulatory landscape, fostering accountability and transparency in AI operations. By adhering to compliance standards, organizations not only avoid legal penalties but also demonstrate their commitment to ethical innovation, which is vital for long-term success and societal acceptance of AI-driven solutions.

AI systems can gather and analyze vast amounts of personal data. It is essential for organizations to ensure this data is obtained and utilized in a legal and ethical manner to avoid potential legal and financial risks.

AI SYSTEMS AND COMPLIANCE WITH THE ARTIFICIAL INTELLIGENCE ACT OF THE EUROPEAN UNION

A survey conducted by Accenture of 850 C-suite executives across 17 geographies and 20 industries to understand their attitudes toward AI regulation and assess their readiness to embrace it disclosed that only about half (47%) of the surveyed organizations have developed a risk management framework, even though it is a prerequisite to developing AI responsibly. The proposed EU AI Act defines different categories of AI risk primarily driven by use case or context. So, a risk management framework will likely become necessary to comply with the regulation once adopted.[3]

WHAT IS THE ARTIFICIAL INTELLIGENCE ACT OF THE EUROPEAN UNION (EU AI ACT)?

The EU AI Act is a proposed regulation by the European Union aimed at ensuring AI systems are safe and ethical and respect fundamental rights. It classifies AI systems into risk categories – unacceptable, high, limited, and minimal – and imposes stricter rules on higher-risk applications, like healthcare or law enforcement. The goal is to balance innovation with protecting people from potential harm caused by AI.

An AI system, according to the EU AI Act, is defined as a machine-based system designed to operate with varying levels of autonomy that may exhibit adaptiveness after deployment and that, for explicit or implicit objectives, infers from the input it receives, how to generate outputs such as predictions, content, recommendations, or decisions that can influence physical or virtual environments.

The EU AI Act includes specific provisions for General Purpose AI (GPAI), recognizing its broad applicability across many sectors. It requires GPAI developers to ensure compliance with rules like transparency, safety, and risk management, especially if their systems are used in high-risk applications. This approach holds developers accountable while aiming to prevent the misuse of GPAI technologies.

GENERAL PURPOSE AI (GPAI)

GPAI model means an AI model, including when trained with a large amount of data using self-supervision at scale that displays significant generality and is capable of competently performing a wide range of distinct tasks regardless of the way the model is placed on the market and that can be integrated into a variety of downstream systems or applications. This does not cover AI models that are used before release on the market for research, development, and prototyping activities.

GPAI system means an AI system that is based on a GPAI model that has the capability to serve a variety of purposes, both for direct use as well as for integration in other AI systems.

GPAI models present systemic risks when the cumulative amount of computing used for its training is greater than 10^{25} floating point operations.

All providers of GPAI models must:

- Draw up technical documentation, including training and testing process and evaluation results.
- Draw up information and documentation to supply to downstream providers that intend to integrate the GPAI model into their AI system so that the latter understands capabilities and limitations and can comply.
- Establish a policy to respect the EU Copyright Directive.
- Publish a sufficiently detailed summary of the content for training the GPAI model.[4]

AI SYSTEMS COMPLIANCE WITH THE EU GENERAL DATA PROTECTION REGULATION

The EU GDPR sets strict requirements for AI systems, especially regarding the processing of personal data. It mandates transparency, accountability, and fairness, requiring organizations to inform users about how their data is processed and to obtain their explicit consent for specific uses.

AI systems must also implement data protection principles, such as minimizing data use, ensuring data accuracy, and enabling users to access, rectify, or delete their data while conducting data protection impact assessments (DPIAs) for high-risk processing activities.

Specific sections of the EU GDPR address compliance requirements relevant to AI systems.

- Article 5 – Principles Relating to Processing of Personal Data

  Requires personal data to be processed lawfully, fairly, and transparently, with collection limited to specific, explicit, and legitimate purposes. Data must be adequate, relevant, and minimized, ensuring accuracy and timely rectification or erasure of inaccuracies. Storage is restricted to the necessary duration, except for certain archiving, research, or statistical purposes with safeguards in place. Data processing must ensure security and protection against unauthorized access, loss, or damage through appropriate technical and organizational measures.
- Article 6 – Lawfulness of Processing

  Processing personal data under the GDPR is lawful only if it meets specific conditions, such as obtaining the data subject's consent, fulfilling a contractual or legal obligation, protecting vital interests, performing a task in the public interest, or serving legitimate interests that do not override individual rights. Processing must align with laws set by the EU or Member States, which may define specific requirements, including purpose, data types, storage periods, and safeguards like encryption or pseudonymization. For further processing of data beyond its original purpose, compatibility must be assessed, considering the initial purpose, the relationship between the data subject and controller, and potential consequences for individuals. These provisions ensure AI systems operate legally and ethically when handling personal data.
- Article 22 – Automated Individual Decision-Making, Including Profiling

  This Article gives EU citizens the right not to be subject to decisions based solely on automated processing, including profiling, that have legal or significant effects on them, except in cases where such processing is necessary for a contract, authorized by law with safeguards, or based on explicit consent. Even in these exceptions, safeguards must protect individuals' rights, including the right to human intervention and the ability to express their views and contest decisions. Automated decisions cannot use sensitive personal data unless specific conditions are met with appropriate safeguards in place. These rules ensure transparency and fairness in AI-driven decision-making.

- Article 35 – Data Protection Impact Assessment (DPIA)
  DPIA is required when processing, particularly using new technologies, poses a high risk to individuals' rights and freedoms, such as automated profiling, large-scale processing of sensitive data, or systematic monitoring. The assessment must include a description of processing operations, purposes, risks, and safeguards, ensuring proportionality and compliance with the GDPR. Controllers should seek advice from a Data Protection Officer and, where appropriate, consult data subjects or representatives. Supervisory authorities provide lists of processing activities that require or exempt a DPIA, and assessments must be reviewed when risks change. These steps ensure responsible AI system deployment concerning data protection.[5]

G7 AI GOVERNANCE INITIATIVE

The Hiroshima AI Process is an initiative launched by the G7 in May 2023 to establish inclusive global governance for AI, particularly focusing on advanced AI systems like generative AI. Its primary objective is to maximize AI's innovative opportunities while mitigating associated risks and challenges. The process led to the development of the "Hiroshima AI Process Comprehensive Policy Framework."[6]

International Guiding Principles on AI and a voluntary Code of Conduct for AI developers under the Hiroshima AI Process. The International Code of Conduct for Organizations Developing Advanced AI Systems aims to promote safe, secure, and trustworthy AI worldwide and will provide voluntary guidance for actions by organizations developing the most advanced AI systems, including the most advanced foundation models and generative AI systems.[7]

CHINA'S INTERIM MEASURES FOR THE MANAGEMENT OF GENERATIVE ARTIFICIAL INTELLIGENCE SERVICES

Introduced in 2023, the Measure establishes guidelines to regulate the development and deployment of generative AI technologies, ensuring their safe and ethical use. These measures require companies providing AI services to comply with national laws, prioritize content accuracy, and prevent the dissemination of harmful or false information. Providers must register their generative AI systems with regulatory authorities, perform security assessments, and implement mechanisms to monitor and address potential risks associated with their systems.

The measures also emphasize transparency and user rights, mandating that providers inform users when AI-generated content is involved and provide users the ability to report misuse or inaccuracies. Additionally, companies are held accountable for training datasets, ensuring they meet standards for data quality, legality, and the exclusion of biased or harmful information.

AI AND THE INTERNATIONAL ORGANIZATION FOR STANDARDIZATION (ISO)

The ISO's Role in AI Compliance

The ISO plays a critical role in AI compliance by developing global standards that promote safe, ethical, and responsible development and deployment of AI systems.

Through its subcommittee ISO/IEC JTC 1/SC 42, which focuses on AI, the ISO provides guidelines and frameworks for AI governance, risk management, and technical evaluation.

By creating these internationally recognized benchmarks, the ISO helps organizations across industries align their AI operations with best practices, ensuring regulatory compliance and fostering public trust. The standards also facilitate cross-border collaboration by harmonizing expectations, allowing businesses to innovate responsibly while minimizing risks associated with AI technologies.

ISO STANDARDS FOR AI

The following key standards aim to establish common practices for transparency, fairness, and accountability in AI systems. Auditors and risk assessors may refer to these standards as additional resources for developing audit/assessment questions regarding AI's compliance as defined at the beginning of this chapter.

- **ISO/IEC 5338:2023** Artificial intelligence – AI system lifecycle processes

  This ISO standard defines a set of processes and associated concepts for describing the lifecycle of AI systems based on machine-learning and heuristic systems. This document provides processes that support the definition, control, management, execution, and improvement of the AI system in its lifecycle stages. These processes can also be used within an organization or a project when developing or acquiring AI systems.[8]
- **ISO/IEC 23894:2023** Artificial Intelligence – Guidance on Risk Management

  This document provides guidance on how organizations that develop, produce, deploy, or use products, systems, and services that utilize AI can manage risk specifically related to AI. The guidance also aims to assist organizations in integrating risk management into their

AI-related activities and functions. It moreover describes processes for the effective implementation and integration of AI risk management.[9]

- **ISO/IEC TR 24028:2020** Artificial Intelligence – Overview of Trust worthiness in AI
  This document surveys topics related to trustworthiness in AI systems, including the following:
  - Approaches to establishing trust in AI systems through transparency, explainability, controllability, etc.
  - Engineering pitfalls and typical associated threats and risks to AI systems, along with possible mitigation techniques and methods.
  - Approaches to assess and achieve availability, resiliency, reliability, accuracy, safety, security, and privacy of AI systems.
    The specification of levels of trustworthiness for AI systems is out of the scope of this document.[10]
- **ISO/IEC 42001:2023** Artificial Intelligence Management System Standard (AIMS)
  It is an international standard that specifies requirements for establishing, implementing, maintaining, and continually improving an AIMS within organizations. It is designed for entities providing or utilizing AI-based products or services, ensuring responsible development and use of AI systems. The Standard addresses the unique challenges AI poses, such as ethical considerations, transparency, and continuous learning. For organizations, it sets out a structured way to manage risks and opportunities associated with AI, thereby balancing innovation with governance.[11]

AI SYSTEMS COMPLIANCE – THE UNITED STATES

As of December 2024, the United States lacks comprehensive federal legislation specifically targeting AI system compliance.

However, existing laws and regulatory bodies addressed certain aspects of AI compliance:

On October 30, 2023, President Biden issued Executive Order 14110 titled "Safe, Secure, and Trustworthy Development and Use of Artificial Intelligence," establishing comprehensive guidelines to ensure AI technologies are developed and utilized responsibly.

SAFE, SECURE, AND TRUSTWORTHY DEVELOPMENT AND USE OF AI

Key Points on AI Compliance Protocols, Policies, and Procedures:

- Safety and Security Measures: The order mandates rigorous testing and evaluation of AI systems to confirm their safety, reliability, and resilience against misuse. This includes implementing standardized assessments and continuous monitoring to ensure AI functions as intended and complies with federal laws.

- Privacy Protection: Agencies are instructed to address privacy risks potentially exacerbated by AI, including through the facilitation of mass surveillance and the aggregation of personal data. This involves developing policies to safeguard individual privacy rights in AI applications.
- Equity and Civil Rights: The Executive Order emphasizes that AI must advance equity and civil rights and cannot disadvantage marginalized groups. This includes ensuring that AI systems do not exacerbate societal harms such as fraud, discrimination, bias, and disinformation.
- Transparency and Public Awareness: The order calls for the development of effective labeling and content provenance mechanisms so that Americans are able to determine when content is generated using AI and when it is not.
- Federal Leadership and Coordination: The order establishes a coordinated, federal government-wide approach to AI governance, including the appointment of Chief Artificial Intelligence Officers in major federal agencies to oversee AI-related initiatives and ensure adherence to the outlined principles.

This Executive Order represents a significant step by the US government to balance the rapid advancement of AI technologies with the necessity of implementing robust compliance protocols, ensuring that AI development and deployment are conducted in a manner that is safe, secure, and trustworthy.[12]

US FEDERAL TRADE COMMISSION'S (FTC) OPERATION AI COMPLY

On September 25, 2024, the US FTC launched "Operation AI Comply," a law enforcement initiative targeting deceptive practices involving AI.

The operation aims to identify and penalize companies that misuse AI to deceive or defraud consumers, ensuring compliance with existing laws prohibiting unfair or deceptive practices.

Implications for AI Compliance

- Truthful Marketing: Companies must ensure that AI-related claims are accurate and substantiated, avoiding exaggerations about capabilities or outcomes.
- Avoiding Deceptive Practices: Utilizing AI tools to create deceptive content, such as fake reviews, is prohibited and subject to enforcement actions.

- Regulatory Oversight: The FTC is actively monitoring AI-related activities, reinforcing the need for businesses to adhere to fair practices and consumer protection laws.[13]

NATIONAL TELECOMMUNICATIONS AND INFORMATION ADMINISTRATION (NTIA)

In June 2023, 23 Attorneys General responded to the NTIA request for comment on AI policies, specifically commenting on AI system accountability measures and policies, Docket Number NTIA–2023–0005, 88 FR 22433 (July 12, 2022).

The Attorneys General identified four critical areas related to AI compliance:

1. Transparency and Independent Standards. The Attorneys General recommended mandating clear public disclosure of AI's role in decision-making processes, especially for high-risk applications. They proposed that organizations publish policies explaining how AI decisions are made, the role of human oversight, and mechanisms for individuals to appeal AI-driven decisions. They emphasized the need for consistent standards for testing, assessment, and external audits of AI systems, potentially overseen by entities like the NTIA or NIST.
2. Risk-Based Approach. They advocated for a regulatory framework that evaluates AI systems based on their potential impacts, such as risks to civil rights, personal safety, or democratic participation. They suggested focusing on high-risk use cases, like healthcare or automated decision-making with legal or financial implications, while allowing flexibility for low-risk applications.
3. Periodic Testing and Audits. Regular internal and external evaluations of AI systems are deemed essential to identify unintended consequences, such as bias or inaccuracies, and to ensure compliance with established standards. The Attorneys General suggested adopting practices similar to those in data privacy laws, such as Colorado's Privacy Act, which requires data protection assessments for high-risk processing activities.
4. Legislation and Enforcement. The Attorneys General supported federal AI legislation that aligns with existing state privacy laws, focusing on high-risk systems and requiring robust transparency, human oversight, and avenues for individuals to appeal AI-driven decisions. They stressed the importance of concurrent enforcement authority for state Attorneys General to address consumer complaints effectively and ensure accountability across the AI development and deployment chain.[14]

DEPARTMENT OF JUSTICE EVALUATION OF CORPORATE COMPLIANCE PROGRAMS

In September 2024, the Department of Justice (DOJ), via its updated Evaluation of Corporate Compliance Programs, provided guidance to prosecutors when assessing a company's compliance program when an organization comes under investigation for bribery, fraud, or other criminal offenses.

This updated guidance reflects the DOJ's focus on ensuring that compliance programs, including those managing AI, are proactive, adaptable, and integrated into broader risk management strategies.

Below is a summary of the most relevant points from the updated guidance regarding AI compliance:

1. Well-Designed Compliance Programs. The DOJ emphasizes the importance of a compliance program tailored to the company's specific risks, including risks associated with new technologies like AI. Companies are expected to integrate policies and procedures that address risks from AI, ensure proper training for employees, and periodically update compliance programs based on lessons learned and evolving risks. A risk-based approach to resource allocation is encouraged, with more scrutiny applied to high-risk areas.
2. Autonomy, Resources, and Implementation. Effective implementation requires sufficient resources, qualified personnel, and independent oversight. Companies must allocate adequate funding, staff, and tools for compliance functions, including mechanisms for monitoring AI's impact. Compliance teams must have direct access to decision-makers and be empowered to enforce policies. The DOJ also highlights the need for integrating compliance efforts into AI system governance and ensuring controls are in place to monitor AI's reliability and compliance with applicable laws.
3. Accountability and Continuous Improvement. The DOJ advises companies to conduct regular assessments of AI systems, ensure accountability for AI-related decisions, and address potential misuse or unintended consequences. Continuous improvement, periodic testing, and reviews are necessary to maintain effective compliance programs. Companies should also implement clear consequences for non-compliance, such as financial penalties, and incentivize ethical behavior through compensation structures.
4. Investigations and Reporting: A robust compliance program includes mechanisms for employees to report potential violations confidentially and without fear of retaliation. Investigations of AI-related

misconduct should be independent, thorough, and timely, with findings used to identify root causes and improve systems. Companies are expected to monitor AI systems to detect errors or biases early and take corrective actions swiftly.[15]

ADDITIONAL US AI-CENTRIC LEGISLATION

While not necessarily focusing on AI compliance as the previous legislation discussed, the following legislative Acts establish the basis and authority for AI guidance in development, design, and implementation.

- **The AI in Government Act of 2020**
  Establishes an AI Center of Excellence (AI CoE) within the General Services Administration to improve the federal government's ability to adopt and manage AI technologies effectively. The AI CoE is tasked with facilitating the integration of AI across federal agencies, enhancing AI competency, and promoting transparency, fairness, and productivity in government operations. It provides guidance on AI acquisition, data management, and policies to mitigate risks such as bias and discrimination. Additionally, the Act requires the development of workforce planning for AI-related roles, ensuring agencies have the necessary skills and competencies to support AI adoption while maintaining public trust through transparency and civil rights protections.
  [www.congress.gov/116/bills/hr2575/BILLS-116hr2575pcs.pdf]
- **The Advancing American AI Act**
  Promotes the responsible development and adoption of AI within federal agencies, aligning with US values such as privacy, civil rights, and civil liberties. It emphasizes modernizing government systems by integrating AI technologies to enhance efficiency, productivity, and interagency collaboration. Key provisions include creating AI inventories, implementing robust risk management strategies, and ensuring transparency and accountability in AI applications. The Act also fosters innovation by encouraging partnerships between government, academia, and the private sector while addressing challenges like data security, bias mitigation, and workforce upskilling.
  [www.congress.gov/117/bills/s1353/BILLS-117s1353rs.pdf]
- **Blueprint for an AI Bill of Rights 2022**
  Outlines five guiding principles for the responsible design, use, and deployment of automated systems to protect civil rights, privacy, and democratic values. The framework emphasizes protections against unsafe or ineffective systems, algorithmic discrimination, abusive

data practices, lack of transparency, and insufficient human oversight. It advocates for equity in AI systems, privacy by design, clear notice and explanations for users, and the availability of human alternatives or remedies for adverse outcomes.

This non-binding framework serves as a guide for federal agencies, private organizations, and communities, proposing actions like independent evaluations, risk assessments, and proactive equity measures to prevent harm. While not legally enforceable, the Blueprint complements existing laws and policies, providing a vision for protecting individuals and communities from the potential risks of AI technologies while fostering innovation and public trust.

[https://bidenwhitehouse.archives.gov/ostp/ai-bill-of-rights/]

STATE-LEVEL AI COMPLIANCE LEGISLATION

As of December 2024, several US states had enacted legislation directly addressing the development, deployment, and compliance of AI systems.

While many of the states in the United States have begun to design and promote legislation focused on control and oversight of AI, these efforts, for some, are still embryonic and, for others, have not yet been officially passed into law. Table 15.1 summarizes the current laws across the United States addressing AI, as of this writing.

Table 15.1 AI Legislation at the State Level in the United States

State	*AI Legislation Description*
California	In 2024, California introduced the Safe and Secure Innovation for Frontier Artificial Intelligence Models Act (SB 1047), aiming to mandate safety tests for advanced AI models and establish a publicly funded cloud computing cluster. However, Governor Gavin Newsom vetoed the bill in September 2024, expressing concerns about its potential impact on innovation. In September 2024, California passed more than a dozen additional AI-related bills, including measures imposing disclosure requirements on AI companies regarding the datasets used to train their models, requiring healthcare providers to disclose when they use generative AI to communicate with patients, and imposing limits on how healthcare providers and insurers can automate their services.[16]
Colorado	Colorado enacted a comprehensive AI framework (Colorado's Artificial Intelligence Act [CAIA]) in June 2024, effective February 1, 2026, regulating the development, deployment, and use of certain AI systems. The law requires AI developers and deployers to make specific disclosures and documentation available to the state attorney general and the public, focusing on high-risk AI systems.

(*Continued*)

Table 15.1 (Continued)

State	*AI Legislation Description*
New York	New York City's Bias Audit Law (Local Law 144), effective July 5, 2023, prohibits the use of automated employment decision tools unless they have been independently audited for bias. This law aims to prevent discrimination in hiring processes facilitated by AI systems.
Tennessee	In March 2024, Tennessee passed the Ensuring Likeness Image and Voice Security (ELVIS) Act, the first US legislation aimed at regulating AI simulation of image, voice, and likeness. The law prohibits the unauthorized use of AI to clone an individual's voice or likeness without consent, addressing concerns over deepfakes and voice cloning.
Utah	Utah's Artificial Intelligence Policy Act (SB 149), signed into law in March 2024 and effective May 1, 2024, establishes liability for companies that fail to disclose their use of generative AI when required by state consumer protection laws. It also creates the Office of Artificial Intelligence Policy and the Artificial Intelligence Learning Laboratory Program.

US SECTOR-SPECIFIC LEGISLATION AND AI COMPLIANCE

While the United States lacks a comprehensive federal AI regulation, sector-specific legislations, such as the legislations listed here, play a critical role in directing AI compliance.

1. Health Insurance Portability and Accountability Act (HIPAA)
   HIPAA regulates the use and disclosure of protected health information (PHI). AI systems processing PHI must ensure compliance with HIPAA's privacy and security rules, including data encryption, patient consent, and safeguards against unauthorized access.
2. Fair Credit Reporting Act (FCRA)
   FCRA governs how consumer credit information is collected, accessed, and used. AI systems used in credit-scoring or lending decisions must adhere to FCRA guidelines, ensuring accuracy, transparency, and the ability for consumers to dispute inaccurate data.
3. Children's Online Privacy Protection Act (COPPA)
   COPPA regulates the collection and use of personal information from children under 13. AI systems targeting children must obtain verifiable parental consent and comply with stringent data privacy requirements.
4. Equal Credit Opportunity Act (ECOA)
   ECOA prohibits discrimination in credit transactions based on factors like race, gender, or religion. AI systems involved in credit

decisions must be free from bias and discrimination, adhering to fairness and equity principles.

5. Americans with Disabilities Act (ADA)
   The ADA ensures equal access to services for individuals with disabilities. AI technologies, including chatbots or accessibility tools, must accommodate disabled users and ensure compliance with ADA requirements.
6. Gramm-Leach-Bliley Act (GLBA)
   GLBA mandates financial institutions to protect consumer financial data. AI applications in banking must comply with privacy safeguards and inform customers about data-sharing practices.

US PRIVATE SECTOR STANDARDS, FRAMEWORKS, AND DIRECTIVES THAT GUIDE AI COMPLIANCE

1. Microsoft – Responsible AI Standard (v2)
   Microsoft's framework outlines fairness, transparency, reliability, safety, privacy, and accountability principles. Updated in 2022, it provides tools and resources for developers to build and deploy AI systems responsibly, including detailed impact assessments and audits. [https://blogs.microsoft.com/wp-content/uploads/prod/sites/5/2022/06/Microsoft-Responsible-AI-Standard-v2-General-Requirements-3.pdf.]
2. Google's AI Principles (Updated 2023)
   Google emphasizes ethical AI development through commitments to societal benefit, avoiding harm, ensuring fairness, and building explainable systems. Their principles discourage AI use for surveillance and weaponization and emphasize responsible AI research and development. [https://ai.google/static/documents/ai-principles-2023-progress-update.pdf.]
3. IBM – AI Ethics and Trust Framework
   IBM focuses on trustworthiness, transparency, and fairness in AI applications. Its framework emphasizes the need for explainability, user trust, and eliminating bias, alongside detailed guidelines for creating robust governance systems.
   [www.ibm.com/think/insights/a-look-into-ibms-ai-ethics-governance-framework]
4. The Partnership on AI (PAI)
   This industry consortium, including members like Amazon, Microsoft, and IBM, promotes ethical AI development through guidelines that focus on fairness, accountability, transparency, and minimizing societal harm.
   [https://partnershiponai.org]

5. OpenAI – AI Safety and Alignment Standards
   OpenAI provides guidance on ensuring that advanced AI models align with human values and benefit humanity. It emphasizes responsible deployment, research transparency, and collaborations to prevent misuse.
   [https://openai.com/index/openai-safety-update]
6. Responsible AI Licenses (RAIL)
   RAIL initiatives create licenses restricting the use of AI technologies to ensure they are deployed ethically and do not contribute to harm or abuse. These licenses are aimed at open-source AI projects.
   [www.licenses.ai]

AI COMPLIANCE – SECTOR IMPACT

While AI compliance is crucial across all organizations, compliance is deemed most critical in the following industry sectors:

- Healthcare, where privacy and patient safety must adhere to regulations like HIPAA.
- Finance, where transparency and fairness are required under laws like GLB, FCRA, and SOX.
- Education, to ensure fairness and prevent bias in AI-powered assessments and privacy of student information Family Educational Rights Privacy Act (FERPA).
- Employment and hiring, where AI tools must comply with anti-discrimination laws (Title VII of the US Civil Rights Act) and government and public services, where ethical use of AI directly impacts civil liberties and public trust.

GOVERNANCE IN AI COMPLIANCE

Governance refers to the framework of rules, practices, processes, and structures through which an organization is directed and controlled to achieve its objectives, ensure accountability, and manage risks.

The role of governance in AI compliance is to provide a structured framework of accountability, oversight, and decision-making to ensure that AI systems are developed and deployed responsibly and ethically.

Governance establishes the policies, procedures, and mechanisms that guide organizations in aligning AI practices with legal, regulatory, and ethical standards. This includes setting clear accountability for compliance, implementing oversight structures, and ensuring that all actions related to AI adhere to the principles of transparency, fairness, and safety.

Governance also facilitates risk management by identifying potential ethical, legal, and security challenges associated with AI and ensuring that robust cybersecurity measures are in place to mitigate risks. It promotes stakeholder trust by emphasizing transparency in AI operations and decision-making, ensuring that these systems operate reliably and in line with organizational values.

Ultimately, governance in AI compliance ensures that AI systems not only meet legal requirements but also uphold ethical standards, fostering public trust and confidence in their use.

SMART PRACTICES FOR AI COMPLIANCE

A review of the current literature from both researchers and practitioners identifies "smart practices" for achieving AI compliance. The common consensus of "smart practices" for achieving AI compliance includes the following:

1. Data Protection and Privacy
   Implement strong measures to safeguard sensitive and personal data used in AI systems. This includes encryption, anonymization, and ensuring compliance with data protection regulations like GDPR or CCPA.
2. Bias Detection and Mitigation
   Regularly assess AI systems for biases in training data, algorithms, or outputs, and implement fairness strategies to mitigate discriminatory outcomes.
3. Transparent Operations
   Ensure AI systems are explainable and provide clear documentation on how decisions are made, in compliance with regulatory requirements for transparency.
4. Human Oversight and Accountability
   Incorporate human oversight in critical decision-making processes and assign clear accountability for AI-related decisions and outcomes.
5. Risk Assessment and Management
   Conduct comprehensive risk assessments to identify potential threats to safety, privacy, and fairness, and develop robust risVk-mitigation plans.
6. Regular Audits and Monitoring
   Perform routine internal and external audits of AI systems to ensure ongoing compliance with ethical, technical, and legal standards.
7. Ethical Governance Frameworks
   Establish ethical review boards or committees to oversee AI development and deployment, ensuring alignment with societal values and human rights.

8. Security Measures
   Protect AI systems from adversarial attacks and vulnerabilities through secure coding practices, robust validation techniques, and threat detection systems.
9. Employee Training and Awareness
   Educate employees on AI compliance standards, data protection practices, and ethical AI development, and ensure continuous updates on regulatory changes.
10. Collaboration with Legal and Industry Experts
   Work closely with legal advisors and industry specialists to stay informed about evolving regulations and implement best practices effectively.

Bonus

11. Comprehensive Compliance Program
   Develop and maintain a centralized compliance program to oversee all aspects of AI design, development, deployment, and monitoring. This program should integrate ethical, legal, and technical standards across departments, ensuring consistent adherence to policies and fostering accountability throughout the organization.
12. Robust Documentation Practices
   Implement thorough documentation procedures for all stages of the AI lifecycle, including data collection, model development, training, testing, deployment, and updates. Clear, detailed records ensure transparency, facilitate audits, and provide a reference for addressing compliance concerns or technical issues efficiently.

Auditing AI Compliance

The following questions are designed for auditors and risk assessors tasked with examining the internal controls present (or not) in an organization's ability to ensure their AI systems (in production and under development) align with relevant laws, regulations, and ethical standards governing AI development and deployment.

1. How does the organization assess and manage compliance risks associated with third-party AI vendors or partners?
2. How does the company ensure that AI systems are explainable to end-users and decision-makers?
3. How does the organization ensure that all AI systems comply with applicable laws such as GDPR, HIPAA, or industry-specific regulations?
4. How does the organization stay informed about and adapt to changes in AI-related laws and regulations?
5. How does an organization's ethical guidelines influence the design, deployment, and operation of AI systems?

6. How are risks associated with AI systems (e.g., data breaches, bias, incorrect predictions) identified, assessed, and managed?
7. How does the organization ensure that the processes and outcomes of the in-production AI systems are explainable and understandable to all stakeholders?
8. How does the organization assess and address risks associated with the use of AI technologies?
9. How does the organization identify and mitigate biases in AI algorithms and training datasets?
10. How does the organization ensure that AI systems only use data that is strictly necessary for their intended purpose?
11. How does the organization ensure appropriate levels of human oversight in AI-driven decision-making?
12. How does the organization ensure that data used in AI systems is accurate, secure, and compliant with privacy regulations?
13. How does the organization handle incidents related to AI compliance, such as breaches or ethical violations?
14. How does the organization measure and evaluate the performance and compliance of AI systems?
15. How does the organization safeguard AI systems against malicious attacks, including model inversion and adversarial inputs?
16. How does the organization structure its governance framework to ensure AI systems align with relevant laws and ethical standards?
17. How does the organization manage compliance when transferring data used in AI systems across international borders?
18. How often are AI systems and their compliance measures reviewed and updated to ensure continued alignment with evolving laws and standards?
19. What cybersecurity measures are in place to protect AI systems from exploitation or misuse?
20. What ethical principles guide the development and deployment of the organization's AI systems?
21. What systems are in place to create and maintain a comprehensive audit trail for decisions made by AI?
22. What internal policies and procedures have been implemented to guide the responsible development and deployment of AI systems?
23. What measures are in place to ensure that AI decision-making processes are transparent to stakeholders?
24. What processes are in place for the continuous monitoring and auditing of AI systems to ensure compliance?
25. What steps are being taken to prepare for emerging AI compliance challenges and new technologies?
26. What steps are taken to build and maintain trust with stakeholders regarding the use of AI?

27. What strategies does the organization employ to identify and mitigate potential biases in AI models and algorithms?
28. What training programs or initiatives are in place to educate staff about AI compliance and ethical considerations?
29. What ongoing monitoring processes are implemented to ensure AI systems remain compliant and perform as expected?
30. What processes ensure human intervention in decisions made by AI systems, especially in critical or high-risk areas?
31. What is your organization's protocol for responding to incidents involving AI systems, such as erroneous or harmful decisions?
32. Who oversees AI compliance within the organization, and how is accountability enforced?

Auditing Compliance and Regulatory Requirements

While not the primary focus of this chapter, AI is certainly used by organizations to assist in compliance with international, federal, state, and local legislation.

The following questions focus on how organizations may use AI in achieving compliance (financial, legislatively, etc.) as defined at the beginning of this chapter.

1. How does your organization ensure that the AI system complies with all relevant data protection and privacy laws?
2. What processes are in place to monitor changes in AI-specific regulations, and how are these updates integrated into the AI system?
3. Can you describe the steps taken to verify that the AI system adheres to industry standards and best practices?
4. How does your organization ensure that third-party vendors involved in AI development or deployment are compliant with applicable laws and regulations?
5. What internal controls are established to manage the risk of non-compliance with AI-related regulations?
6. How is compliance with AI regulations documented and reported within the organization?
7. What measures are in place to ensure that the AI system's data-processing activities comply with GDPR or other relevant data protection frameworks?
8. Can you provide examples of how your organization has addressed specific regulatory challenges related to AI?
9. How does your organization audit the AI system to ensure ongoing compliance with evolving legal and regulatory requirements?
10. What contingency plans are in place if the AI system is found to be non-compliant with regulatory requirements?

NOTES

1 McGrath, A., and Jonker, A., October 4, 2024, "AI Compliance: What it is, why it Matters and How to Get Started," IBM, www.ibm.com/think/insights/ai-compliance#:~:text=AI%20compliance%20refers%20to%20the,models%20and%20their%20algorithms%20responsibly.

2 KPMG, January 2024, "KPMG Generative AI Consumer Trust Survey," https://kpmg.com/kpmg-us/content/dam/kpmg/corporate-communications/pdf/2024/kpmg-generative-ai-consumer-trust-survey.pdf.

3 Eitel-Porter, R., and Grosskopf, E., et al., 2022, "From AI compliance to competitive advantage, Becoming responsible by design," Accenture, www.accenture.com/content/dam/accenture/final/a-com-migration/r3-3/pdf/pdf-179/accenture-responsible-by-design-report.pdf.

4 (n.a.), May 30, 2024, "High-level summary of the AI Act," The EU Artificial Intelligence Act, https://artificialintelligenceact.eu/high-level-summary.

5 Marcella, A., December 2024, author summaries of GDPR Articles 5, 6, 22, and 35 using material from the Official Journal of the European Union, May 4, 2016, "Regulation (EU) 2016/679 Of The European Parliament and of The Council of 27 April 2016 on the protection of natural persons with regard to the processing of personal data and on the free movement of such data, and repealing Directive 95/46/EC (General Data Protection Regulation), http://eur-lex.europa.eu/legal-content/EN/TXT/PDF/?uri=CELEX:32016R0679.

6 (n.a.), (n.d.), "The Hiroshima AI Process, Documents of Achievement," www.soumu.go.jp/hiroshimaaiprocess/en/documents.html?utm_source=chatgpt.com.

7 (n.a.), October 30, 2023, "Hiroshima Process International Code of Conduct for Advanced AI Systems, Policy and Legislation," https://digital-strategy.ec.europa.eu/en/library/hiroshima-process-international-code-conduct-advanced-ai-systems.

8 ISO, December 2023, "ISO/IEC 5338:2023," The International Organization for Standardization, www.iso.org/standard/81118.html.

9 ISO, February 2023, "Artificial Intelligence - Guidance on Risk Management," The International Organization for Standardization, www.iso.org/standard/77304.html.

10 ISO, May 2020, "Artificial Intelligence — Overview of Trustworthiness in Artificial Intelligence," The International Organization for Standardization, www.iso.org/standard/77608.html.

11 ISO, December 2023, "Artificial Intelligence Management System Standard (AIMS)," The International Organization for Standardization, www.iso.org/standard/81230.html.

12 Executive Office of the President of the United States, November 1, 2023, "Safe, Secure, and Trustworthy Development and Use of Artificial Intelligence, A Presidential Document, www.federalregister.gov/documents/2023/11/01/2023–24283/safe-secure-and-trustworthy-development-and-use-of-artificial-intelligence.

13 FTC, September 25, 2024, "FTC Announces Crackdown on Deceptive AI Claims and Schemes," Federal Trade Commission, https://www.ftc.gov/news-events/news/press-releases/2024/09/ftc-announces-crackdown-deceptive-ai-claims-schemes.

14 Weiser, P., Tong, W., Skrmetti, J., and Miyares, J., et al., June 12, 2023, "Comment on Artificial Intelligence ("AI") system accountability measures and policies—Docket Number NTIA–2023–0005, 88 FR 22433 (July 12, 2022)," https://oag.ca.gov/system/files/attachments/press-docs/NTIA%20AI%20Comment.pdf.
15 US Department of Justice, September 2024, "Evaluation of Corporate Compliance Programs," Criminal Division, www.justice.gov/criminal/criminal-fraud/page/file/937501/dl.
16 Ravi, S., Southwell, A., and Babendreier, T., December 19, 2024, "AI and the Law: Emerging Trends in Enforcement," McDermott Will & Emery, www.mwe.com/insights/ai-and-the-law-emerging-trends-in-enforcement/?utm_source=chatgpt.com.

Part 5

Auditing AI Systems

Chapter 16

Auditing AI System Documentation and Reporting

DEFINING IT SOFTWARE DOCUMENTATION

Software system documentation means the organized collection of documents that describe the requirements, capabilities, limitations, design, operation, and maintenance of a computer program, operating system, or hardware device. Software system documentation includes but is not limited to data dictionaries, system flow charts, and program documentation that describe the inputs, processing, and outputs, query, update, and reporting program of the software system. System software also includes but is not restricted to, micro-code embedded in hardware (i.e., "firmware," operating systems, communications, system and network management, and utility software).[1]

Artificial intelligence (AI) system program documentation involves detailed records that explain the development, functionality, and ethical considerations of AI models and systems.

Calls to standardize and enact documentation of algorithmic harms and impacts are now commonplace. However, documentation standards for AI remain inchoate and fail to match the capabilities and social effects of increasingly impactful architectures such as Large Language Models (LLMs).

The goal of a documentation system for AI should be to take continuous snapshots of a system's performance in a similar way to how model cards (addressed separately in this chapter) can provide a rich picture of a single model's biases and behavior at a single point in time.[2]

> If you don't know what's in the training data, then you are not positioned to decide if you can safely deploy the thing.[3]

DIFFERENCES BETWEEN TRADITIONAL APPLICATION/SYSTEM DOCUMENTATION AND AI SYSTEM DOCUMENTATION

Traditional software documentation focuses on detailing system architecture, code functionality, user guides, and maintenance procedures.

DOI: 10.1201/9781003646501-21

It emphasizes the static aspects of software systems, providing clear instructions on system operations and interactions.

In contrast, AI system documentation must address additional complexities:

- **Model Behavior and Training**
  AI documentation needs to explain how models are trained, including data sources, training methodologies, and the rationale behind model decisions. This is crucial for understanding and interpreting AI behavior.
- **Ethical and Bias Considerations**
  AI systems can inadvertently perpetuate biases present in training data. Documentation should include information on measures taken to detect and mitigate bias, ensuring ethical AI deployment.
- **Dynamic Learning**
  Unlike traditional systems, some AI models evolve through continuous learning. Documentation must account for these changes, detailing how models adapt and the implications for system behavior.
- **Transparency and Explainability**
  AI systems often operate as "black boxes," making it challenging to interpret their decision-making processes. Comprehensive documentation is essential for transparency and building trust among users and stakeholders.

DOCUMENTING SCOPE, OBJECTIVE, AND INTENDED USE

The specific type and depth of detail required by the documentation users need to prepare will depend highly on the nature of the AI model. The documentation required to substantiate regulatory, compliance, or stakeholder inquiry for a publicly available dataset will be different in its scope and depth than documentation for an LLM used by banks to detect fraud or money-laundering activities or a hospital to diagnose specific types of cancers.

Verifying the scope and objective of the AI system under review is essential. This will assist in determining both the breadth and depth of the documentation required to substantiate the system's operational objectives, as well as to document the development, functionality, and ethical considerations used in the development of the AI system itself.

Additionally, the intended use of the documentation should be determined. Will the documentation be used solely by the organization for the in-house operation, maintenance, and ongoing development of the AI systems? Or will the AI system be public-facing, accessed, and used by various external persons and organizations?

The documentation for an LLM, used by an organization operating in a regulated industry, such as healthcare (HIPAA and FDA), energy and utilities (FERC and EPA), and telecommunications (FCC), will differ from the documentation used in non-regulated industries, (e.g., hospitality and tourism, film production, and gaming).

Determining the scope and objective of the AI system along with its intended use are essential steps in determining specifically what documentation and to what level of detail should that documentation be prepared and available for examination.

Auditing AI System Documentation (General)

The following questions are intended to provide a general insight into the organization's documentation practices related to AI systems under development or in production within the organization.

1. How does the organization ensure that all documentation related to the AI system, such as policies and procedures, are accurate, up-to-date, and complete?
2. Describe the process for creating and maintaining user guides and other instructional materials for the AI system.
3. What measures are in place to verify that the documentation aligns with the actual operation and configuration of the AI system?
4. How does the organization ensure that audit trails are comprehensive, tamper-proof, and readily accessible for review?
5. What internal controls exist to ensure the consistency and accuracy of the reporting mechanisms associated with the AI system? How are these controls documented?
6. How are changes to the AI system documented, and what processes ensure that these updates are communicated effectively to relevant stakeholders?
7. Please provide examples of how documentation has been used to identify and resolve issues within the AI system.
8. What processes are in place to review and approve the documentation and reporting materials related to the AI system?
9. How does the organization ensure that the documentation provides sufficient detail for external audits or regulatory reviews?
10. What feedback mechanisms exist for users or auditors to report deficiencies or areas for improvement in the AI system's documentation?

The remainder of this chapter examines AI documentation artifacts in greater depth.

AUDITING AI DOCUMENTATION ARTIFACTS

WHAT IS AN AI DATASET?

An AI dataset provides the actual data used for training, validating, or testing an AI model.

The foundation of any machine-learning (ML) model is the dataset.

A dataset is a structured collection of data, often labeled or annotated, that is used to train, validate, and test ML models. It serves as the foundation for learning patterns, features, and relationships within the AI system. Dataset documentation should capture aspects like data collection timing, origin, representativeness, and the dataset's purpose. For labeled datasets, the annotation process, quantity of data annotated, validation measures, and annotators' identities are additional components to consider.[4]

Auditing the AI Dataset

1. Does the dataset contain any known biases, and how were they addressed during preparation?
2. How are security measures implemented to prevent unauthorized changes to the dataset?
3. How was the dataset collected, and were ethical and legal guidelines adhered to?
4. Were there any exclusions or omissions in the dataset, and what is the rationale for those?
5. What documentation exists that describes how the dataset was annotated, and who performed the annotations?
6. What evidence exists of dataset validation to ensure accuracy and consistency?
7. What is the size and diversity of the dataset, and how does it impact model generalization?
8. What preprocessing steps (e.g., cleaning, normalization) were applied to the dataset?
9. What purpose was the dataset created for, and how does this align with the AI system's objectives?
10. What versioning controls are in place for updates to the dataset?

DATASHEETS[5]

What Is an AI Datasheet?

An AI datasheet details the dataset's creation, composition, intended use, limitations, and ethical considerations. This ensures transparency, accountability, and proper use of the dataset within AI workflows.

The AI datasheet is, therefore, a structured document that provides comprehensive, standardized information about an AI system or application. It is designed to ensure transparency, accountability, and informed decision-making for stakeholders. This datasheet should detail the system's specifications, training data, operational constraints, ethical considerations, and compliance with relevant regulations.

An AI datasheet is similar to data sheets required in various industries such as (representative list):

Chemical and Manufacturing Industries

- Type of Data Sheet: Safety Data Sheets (SDS) or Material Safety Data Sheets (MSDS)
- Governing Body: OSHA (Occupational Safety and Health Adminis tration), REACH (Registration, Evaluation, Authorization, and Restriction of Chemicals), GHS (Globally Harmonized System of Classification and Labeling of Chemicals)

Automotive Industry

- Type of Data Sheet: Technical Data Sheets (TDS), Material Specification Sheets
- Governing Body: Society of Automotive Engineers, International Organization for Standardization

Energy and Utilities

- Type of Data Sheet: Specification Sheets for Equipment, SDS for Lubricants and Fuels
- Governing Body: ISO, OSHA, NFPA (National Fire Protection Association)

IT and Software

- Type of Data Sheet: Product TDS, System Requirement Sheets
- Governing Body: NIST (National Institute of Standards and Technology), IEEE (Institute of Electrical and Electronics Engineers)

Summary

The relationship between an AI dataset and AI datasheets is that AI datasheets serve as comprehensive documentation for the associated dataset. An AI dataset provides the actual data used for training, validating, or testing an AI model, while the AI datasheet details the dataset's creation, composition, intended use, limitations, and ethical considerations. This ensures transparency, accountability, and proper use of the dataset within AI workflows.

- Datasheet Definition: Documents that detail key attributes and specifications of datasets used for AI model training, validation, and testing. They describe the dataset's purpose, composition, collection methods, and ethical considerations. Datasheets provide a concise, clear, and comprehensive view of an AI model, ensuring transparency and usability for end-users and stakeholders.

- Use in Audit: Evaluates the suitability and quality of datasets, ensuring they align with ethical and regulatory guidelines.
- See Appendix 6 – AI Model Datasheet – Model Name: GPT-4.

Purpose of an AI System Datasheet

An AI datasheet's four primary objectives are:

1. Transparency: To help stakeholders (e.g., users, regulators, developers) understand how the AI system functions, including its strengths and weaknesses.
2. Trust: To foster confidence in the system's design, ethical use, and compliance with regulations.
3. Accountability: To establish clear lines of responsibility and compliance for the system's lifecycle.
4. Standardization: To provide a unified framework for documenting and assessing AI systems, reducing variability and misunderstandings.

This type of AI datasheet aligns with emerging guidelines and frameworks for responsible AI development, such as those from the IEEE, NIST, and OECD AI Principles.

Key Components of an AI System Datasheet

The following are key components that may be derived through the use of AI system datasheets.

System Overview

- Description of the AI system's purpose, capabilities, and intended use cases.
- High-level explanation of the system's architecture and key features.

Training and Dataset Information

- Data sources used for training, validation, and testing.
- Description of dataset size, origin, and any preprocessing techniques.
- Information on potential biases and measures taken to mitigate them.

Performance Metrics

- Quantitative metrics (e.g., accuracy, precision, recall, F1 score).
- Benchmarks for performance under various conditions.

Limitations and Risks

- Known limitations or edge cases where the AI system may fail.

- Potential risks, such as biases, misinterpretation of outputs, or security vulnerabilities.

Ethical Considerations

- Information on fairness, inclusivity, and adherence to ethical principles.
- Measures taken to ensure privacy, avoid harm, and promote accountability.

Compliance and Governance

- Compliance with industry standards and regulatory frameworks (e.g., General Data Protection Regulation (GDPR), ISO standards for AI, NIST AI Risk Management Framework).
- Details of third-party audits or certifications, if applicable.

User and Deployment Information

- Instructions for installation, integration, and use.
- Recommended operational environment and hardware/software dependencies.

Maintenance and Updates

- Process for system updates, retraining, or version control.
- Contact information for support and reporting issues.

The reader is directed to Appendix 12 – AI Application Datasheet Example for the fictitious product "Sentiment Analyzer Pro 2.0," which provides an example of an AI Application Datasheet.

Gebru et al., in their article "Datasheets for Datasets," recognized that despite the importance of data to ML, there was no standardized process for documenting ML datasets.

The authors proposed that every dataset be accompanied by a datasheet that documents its motivation, composition, collection process, and recommended uses...stating that "Data Sheets for datasets have the potential to increase transparency and accountability within the machine learning community, mitigate unwanted societal biases in machine learning models, facilitate greater reproducibility of machine learning results, and help researchers and practitioners to select more appropriate datasets for their chosen tasks."[6]

Since proposing their idea of creating an AI datasheet to accompany AI datasets, organizations such as IBM, Microsoft, and Google have adopted this practice.

Datasheets for datasets propose questions ranging from dataset motivation to maintenance. The following material focuses the reader's attention on auditing AI datasheets for datasets.

Auditing AI Datasheets for AI Datasets

Motivation. The following questions are primarily intended to encourage dataset creators to clearly articulate their reasons for creating the dataset and to promote transparency about funding interests. The latter may be particularly relevant for datasets created for research purposes.

1. For what purpose was the dataset created? Was there a specific task in mind? Was there a specific gap that needed to be filled? Please provide a description.
2. Who created the dataset (e.g., which team, research group) and on behalf of which entity (e.g., company, institution, organization)?
3. Who funded the creation of the dataset? If there is an associated grant, please provide the name of the grantor and the grant name and number.

   Composition. Dataset creators should read through these questions before any data collection and then provide answers once data collection is complete. Most of the questions here are intended to provide dataset consumers with the information they need to make informed decisions about using the dataset for their chosen tasks. Some of the questions are designed to elicit information about compliance with the EU's GDPR or comparable regulations in other jurisdictions.

   Questions that apply only to datasets that relate to people are grouped together at the end of the section. Gebru et al. recommend taking a broad interpretation of whether a dataset relates to people. For example, any dataset containing text that was written by people relates to people.
4. What do the instances that comprise the dataset represent (e.g., documents, photos, people, countries)? Are there multiple types of instances (e.g., movies, users, and ratings; people and interactions between them; nodes and edges)? Please provide a description.
5 How many instances are there in total (of each type, if appropriate)?
6 Does the dataset contain all possible instances or is it a sample (not necessarily random) of instances from a larger set? If the dataset is a sample, then what is the larger set? Is the sample representative of the larger set (e.g., geographic coverage)? If so, please describe how this representativeness was validated/verified. If it is not representative of the larger set, please describe why not (e.g., to cover a more diverse range of instances, because instances were withheld or unavailable).
7. What data does each instance consist of? "Raw" data (e.g., unprocessed text or images) or features? In either case, please provide a description.
8. Is there a label or target associated with each instance? If so, please provide a description.

9. Is any information missing from individual instances? If so, please provide a description, explaining why this information is missing (e.g., because it was unavailable). This does not include intentionally removed information but might include, e.g., redacted text.
10. Are relationships between individual instances made explicit (e.g., users' movie ratings and social network links)? If so, please describe how these relationships are made explicit.
11. Are there recommended data splits (e.g., training, development/validation, testing)? If so, please describe these splits, explaining the rationale behind them.
12. Are there any errors, sources of noise, or redundancies in the dataset? If so, please provide a description.
13. Is the dataset self-contained, or does it link to or otherwise rely on external resources (e.g., websites, tweets, and other datasets)? If it links to or relies on external resources, are there:

    a) Guarantees that they will exist and remain constant over time?
    b) Official archival versions of the complete dataset (i.e., including the external resources as they existed at the time the dataset was created)?
    c) Any restrictions (e.g., licenses, fees) associated with any of the external resources that might apply to a dataset consumer?

    Please provide descriptions of all external resources and any restrictions associated with them, as well as links or other access points, as appropriate.
14. Does the dataset contain data that might be considered confidential (e.g., data that is protected by legal privilege or by doctor–patient confidentiality, data that includes the content of individuals' non-public communications)? If so, please provide a description.
15. Does the dataset contain data that, if viewed directly, might be offensive, insulting, threatening, or might otherwise cause anxiety? If so, please describe why.

    If the dataset does not relate to people, you may skip the remaining questions in this section.
16 Does the dataset identify any subpopulations (e.g., by age or gender)? If so, please describe how these subpopulations are identified and provide a description of their respective distributions within the dataset.
17. Is it possible to identify individuals (i.e., one or more natural persons), either directly or indirectly (i.e., in combination with other data) from the dataset? If so, please describe how.
18. Does the dataset contain data that might be considered sensitive in any way (i.e., data that reveals race or ethnic origins, sexual orientations, religious beliefs, political opinions or union memberships, or locations; financial or health data; biometric or genetic data; forms of government identification, such as social security numbers; criminal history)? If so, please provide a description.

Collection process. As with the questions in the previous section, dataset creators should read through these questions before any data collection to flag potential issues and then provide answers once the collection is complete. In addition to the goals outlined earlier, the following questions are designed to elicit information that may help researchers and practitioners create alternative datasets with similar characteristics. Questions that apply only to datasets that relate to people are grouped together at the end of the section.

19. How was the data associated with each instance acquired? Was the data directly observable (e.g., raw text, movie ratings), reported by subjects (e.g., survey responses), or indirectly inferred/derived from other data (e.g., part-of-speech tags, model-based guesses for age or language)? If the data was reported by subjects or indirectly inferred/derived from other data, was the data validated/verified? If so, please describe how.
20. What mechanisms or procedures were used to collect the data (e.g., hardware apparatuses or sensors, manual human curation, software programs, software APIs)? How were these mechanisms or procedures validated?
21. If the dataset is a sample from a larger set, what was the sampling strategy (e.g., deterministic, probabilistic with specific sampling probabilities)?
22. Who was involved in the data collection process (e.g., students, crowd workers, contractors), and how were they compensated (e.g., how much were workers paid)?
23. Over what time frame was the data collected? Does this time frame match the creation time frame of the data associated with the instances (e.g., the recent crawl of old news articles)? If not, please describe the time frame in which the data associated with the instances was created.
24. Were any ethical review processes conducted (e.g., by an institutional review board)? If so, please describe these review processes, including the outcomes, as well as a link or another access point to any supporting documentation.

If the dataset does not relate to people, you may skip the remaining questions in this section.

25. Did you collect the data from the individuals in question directly, or obtain it via third parties or other sources (e.g., websites)?
26. Were the individuals in question notified about the data collection? If so, please describe (or show with screenshots or other information) how notice was provided, and provide a link or another access point to, or otherwise reproduce, the exact language of the notification itself.
27. Did the individuals in question consent to the collection and use of their data? If so, please describe (or show with screenshots or other

information) how consent was requested and provided, and provide a link or another access point to, or otherwise reproduce, the exact language to which the individuals consented.

28. If consent was obtained, were the consenting individuals provided with a mechanism to revoke their consent in the future or for certain uses? If so, please provide a description, as well as a link or another access point to the mechanism (if appropriate).
29. Has an analysis of the potential impact of the dataset and its use on data subjects (e.g., a data protection impact analysis) been conducted? If so, please provide a description of this analysis, including the outcomes, as well as a link or another access point to any supporting documentation.

Preprocessing/cleaning/labeling. Dataset creators should read through these questions before any preprocessing, cleaning, or labeling and then provide answers once these tasks are complete. The questions in this section are intended to provide dataset consumers with the information they need to determine whether the "raw" data has been processed in ways that are compatible with their chosen tasks. For example, text that has been converted into a "bag of words" is not suitable for tasks involving word order.

30. Was any preprocessing/cleaning/labeling of the data done (e.g., discretization or bucketing, tokenization, part-of-speech tagging, SIFT feature extraction, removal of instances, processing of missing values)? If so, please provide a description. If not, you may skip the remaining questions in this section.

Note to Reader

SIFT stands for Scale-Invariant Feature Transform, which is a computer vision algorithm used to detect and describe local features in images. It is particularly valuable for tasks such as object recognition, image stitching, and 3D reconstruction.

Key Characteristics of SIFT

- Scale-Invariance: SIFT can identify features in an image regardless of the scale of the image, making it robust to zoom in or out.
- Rotation-Invariance: SIFT detects features in an image that remains consistent even if the image is rotated.
- Feature Description: Each detected feature is described using a 128-dimensional vector, which captures information about the gradients around the feature, making it robust to small distortions or noise.
- Applications: Commonly used in image processing tasks, including matching points between images and tracking objects in videos.

In the context of this audit question, SIFT might refer to an example of preprocessing or feature extraction applied to images within a dataset. This step would extract key points from the images for analysis or use as input to a ML algorithm.

31. Was the "raw" data saved in addition to the preprocessed/cleaned/labeled data (e.g., to support unanticipated future uses)? If so, please provide a link or another access point to the "raw" data.
32. Is the software that was used to preprocess/clean/label the data available? If so, please provide a link or another access point.
33. Any other comments?

Uses. The following questions are intended to encourage dataset creators to reflect on the tasks for which the dataset should and should not be used. By explicitly highlighting these tasks, dataset creators can help dataset consumers make informed decisions, thereby avoiding potential risks or harms.

34. Has the dataset been used for any tasks already? If so, please provide a description.
35. Is there a repository that links to any or all papers or systems that use the dataset? If so, please provide a link or another access point.
36. What (other) tasks could the dataset be used for?
37. Is there anything about the composition of the dataset or the way it was collected and preprocessed/cleaned/labeled that might impact future uses? For example, is there anything that a dataset consumer might need to know to avoid uses that could result in unfair treatment of individuals or groups (e.g., stereotyping, quality of service issues) or other risks or harms (e.g., legal risks, financial harms)? If so, please provide a description. Is there anything a dataset consumer could do to mitigate these risks or harms?
38. Are there tasks for which the dataset should not be used? If so, please provide a description.

Distribution. Dataset creators should provide answers to these questions before distributing the dataset either internally within the entity on behalf of which the dataset was created or externally to third parties.

39. Will the dataset be distributed to third parties outside of the entity (e.g., a company, institution, or organization) for whom the dataset was created? If so, please provide a description.
40. How will the dataset be distributed (e.g., tarball on website, API, GitHub)? Does the dataset have a digital object identifier (DOI)?

Note to Reader

A tarball refers to a type of archive file format used primarily on Unix-based systems (such as Linux). It combines multiple files and directories into a single file, typically for easier distribution or storage.

Key Details about Tarballs

1. **File Format:** Tarballs are created using the **tar** (tape archive) command, and the resulting file typically has a.**tar** extension. It may also have extensions like.tar.gz or.tar.bz2 if compressed.
   - **.tar.gz:** A tarball compressed with gzip.
   - **.tar.bz2:** A tarball compressed with bzip2.
2. **Purpose:** Tarballs are often used to bundle datasets, software source code, or other collections of files for downloading or sharing via websites or repositories.
3. **Advantages:**
   - Combines multiple files into one for convenience.
   - Compression reduces file size for quicker downloads.
4. **Example:**
   - A dataset distributed as a tarball might be named dataset.tar.gz, and users would extract it with a command like **tar -xvzf** dataset.tar.gz.

41. When will the dataset be distributed?
42. Will the dataset be distributed under a copyright or other intellectual property (IP) license, and/or under applicable terms of use (ToU)? If so, please describe this license and/or ToU, and provide a link or another access point to, or otherwise reproduce, any relevant licensing terms or ToU, as well as any fees associated with these restrictions.
43. Have any third parties imposed IP-based or other restrictions on the data associated with the instances? If so, please describe these restrictions, and provide a link or another access point to, or otherwise reproduce, any relevant licensing terms, as well as any fees associated with these restrictions.
44. Do any export controls or other regulatory restrictions apply to the dataset or to individual instances? If so, please describe these restrictions, and provide a link or another access point to, or otherwise reproduce, any supporting documentation.

    Maintenance. As with the previous questions, dataset creators should provide answers to these questions before distributing the dataset. The questions in this section are intended to encourage dataset creators to plan for dataset maintenance and communicate this plan to dataset consumers.
45. Who will be supporting/hosting/maintaining the dataset?
46. How can the owner/curator/manager of the dataset be contacted (e.g., email address)?
47. Is there an erratum? If so, please provide a link or another access point.
48. Will the dataset be updated (e.g., to correct labeling errors, add new instances, or delete instances)? If so, please describe how often, by

whom, and how updates will be communicated to dataset consumers (e.g., mailing list, GitHub).

49. If the dataset relates to people, are there applicable limits on the retention of the data associated with the instances (e.g., were the individuals in question told that their data would be retained for a fixed period of time and then deleted)? If so, please describe these limits and explain how they will be enforced.
50. Will older versions of the dataset continue to be supported/hosted/maintained? If so, please describe how. If not, please describe how its obsolescence will be communicated to dataset consumers.
51. If others want to extend/augment/build on/contribute to the dataset, is there a mechanism for them to do so? If so, please provide a description. Will these contributions be validated/verified? If so, please describe how. If not, why not? Is there a process for communicating/distributing these contributions to dataset consumers? If so, please provide a description.

DATA CARDS

What Are Data Cards?

AI data cards are detailed documentation tools that provide comprehensive information about the datasets used to train, test, and validate AI models. These cards include metadata about data sources, preprocessing steps, dataset structure, and potential biases. The goal of AI data cards is to enhance transparency, reproducibility, and accountability in AI development by offering clear insights into the data lifecycle.

A data cards contain a structured collection of summaries gathered over the lifecycle of a dataset about observable (e.g., dataset attributes) and unobservable (e.g., intended use cases) aspects needed for decisions in organizational and practice-oriented contexts. Beyond metadata, data cards include explanations, rationales, and instructions about the provenance, representation, usage, and fairness-informed evaluations of datasets for ML models. Data cards capture critical information about a dataset across its lifecycle. As is true with every dataset, each data card is unique, and no single template satisfactorily captures the nuance of all datasets.[7]

Further, as Pushkarna et al. state, a data card is a dataset documentation framework that increases transparency across dataset lifecycles. Data cards are transparency artifacts that provide structured summaries of ML datasets with explanations of processes and rationale that shape the data and describe how the data may be used to train or evaluate models.

At a minimum, data cards include the following:

1. Upstream sources.
2. Data collection and annotation methods.
3. Training and evaluation methods.
4. Intended use.
5. Decisions affecting model performance.[8]

An AI Data Card is a concise and user-friendly summary designed to quickly convey essential details about an AI system, focusing on accessibility for non-technical audiences.

Key Components of a Data Card

Dataset Overview

- Name, description, scope, and purpose of the dataset.
- Purpose of the dataset and its intended use cases.
- Source and domain of the data (e.g., social media, medical records, images).

Data Collection

- Methods used to collect the data.
- Sources of data.
- Time period and location of data collection.
- Tools or processes used for gathering data.

Data Composition

- Number of samples, classes, or instances.
- Types of data included (e.g., text, images, numerical data).
- Features and labels (if applicable).

Dataset Structure

- Details about the dataset format, features, and labels.

Annotations and Labels

- Information on how data was labeled (e.g., manual labeling, automated tools).
- Qualifications of annotators and guidelines followed during labeling.
- Potential biases introduced during annotation.

Ethical Considerations

- Privacy and consent: Was user consent obtained for data collection?
- Potential risks, such as harm, bias, or misrepresentation.
- Steps taken to mitigate ethical concerns.

Data Quality

- Known errors, inconsistencies, or missing data.
- Preprocessing steps applied (e.g., normalization, tokenization).

Bias and Fairness

- Analysis of potential biases (e.g., demographic representation).
- Limitations of the dataset in terms of diversity or coverage.

Usage Guidelines

- Appropriate and inappropriate uses of the dataset.
- Recommended evaluation metrics and methodologies.
- Known failure modes when using the dataset in AI models.

Preprocessing Steps

- Documentation of any data cleaning, normalization, or transformation processes applied.

Licensing and Access

- Terms of use or licenses governing the dataset.
- Access mechanisms, such as links or repositories.

Maintenance and Updates

- Information on dataset versioning.
- Plans for updates, corrections, or expansions.

Why Are Data Cards Important?

Data cards are important in the overall governance, development, and deployment of AI systems for the following reasons:

1. Transparency: They provide clear documentation about the dataset's origins, structure, and limitations.
2. Trust and Fairness: Users can assess potential biases and limitations before employing the data.
3. Compliance: Data cards help meet ethical AI principles and legal requirements, such as GDPR or other data protection laws.
4. Reproducibility: Detailed information ensures that research using the dataset can be replicated.

The reader unfamiliar with AI data cards is directed to the following websites for additional information and further study:

Data Cards

AI Data Cards – Learning Resource #1

[https://dl.acm.org/doi/fullHtml/10.1145/3531146.3533231#:~:text=Data%20Cards%20are%20structured%20summaries,lifecycle%20for%20responsible%20AI%20development.]

Role of Data Cards
AI Data Cards – Learning Resource #2
[https://datatonic.com/insights/responsible-ai-data-model-cards/]
The Data Cards Playbook
AI Data Cards – Learning Resource #3
[https://sites.research.google/datacardsplaybook/]

Data Cards: Usefulness in Auditing AI Tools

Using AI data cards during an audit can significantly enhance the evaluation process by providing a structured and comprehensive overview of the datasets used in the AI tool.

They are most appropriately used to assess areas related to:

- Data Quality: Ensuring the datasets are accurate, complete, and reliable.
- Bias and Fairness: Identifying and mitigating potential biases in the data.
- Transparency: Understanding the origins and transformations of the data used in the AI tool.

Auditing the AI Tool Using the Tool's AI Data Card

- Dataset Overview
  1. What is the primary purpose and scope of the dataset?
  2. How representative is the dataset of the real-world scenarios it aims to model?

- Data Collection
  1. What are the sources of the data, and how was it collected?
  2. Were there any specific criteria or filters applied during data collection?

- Dataset Structure
  1. What is the structure of the dataset, including features and labels?
  2. Are there any missing or incomplete data points, and how were they handled?

- Preprocessing Steps
  1. What preprocessing steps were applied to the data before using it in the model?
  2. How were outliers, anomalies, and noise in the data addressed?

- Usage Guidelines
  1. What are the recommended use cases for this dataset?
  2. Are there any known limitations or constraints on the dataset's use?

- Ethical Considerations
  1. What potential biases have been identified in the dataset?
  2. What steps have been taken to mitigate these biases and ensure fairness?

- Data Quality
  1. How was the quality of the data assessed and validated?
  2. What measures are in place to ensure the ongoing accuracy and reliability of the data?

- Bias and Fairness
  1. How does the dataset ensure a fair representation of different demographic groups?
  2. Are there any known issues related to bias in the dataset?

- Transparency
  1. How transparent is the documentation regarding the data collection and preprocessing methods?
  2. Are there any undisclosed or unclear aspects of the data lifecycle?

- Compliance and Ethical Use
  1. How does the dataset comply with relevant regulations and standards?
  2. What ethical guidelines were followed in the creation and use of the dataset?

While model cards or system cards focus on AI models and systems, data cards emphasize the datasets powering those models. Together, they form a cohesive framework for AI development and deployment transparency. Table 16.1 summarizes Datasheets vs Data Cards.

Table 16.1 Datasheets vs Data Cards

Characteristic	*AI Datasheets*	*AI Data Cards*
Scope	Comprehensive, covering all aspects of the AI system lifecycle, including data collection, training, testing, and deployment.	Narrower focus, summarizing key highlights and essential facts.
Purpose	To provide in-depth technical and procedural insights for developers, auditors, and regulators.	To offer a quick overview for general understanding and decision-making.

(*Continued*)

Table 16.1 (Continued)

Characteristic	*AI Datasheets*	*AI Data Cards*
Level of Detail	High level of detail with exhaustive descriptions, metrics, and technical specifics.	Low level of detail, emphasizing simplicity and readability.
Audience	Primarily technical stakeholders like engineers, researchers, and regulators.	Broader audience, including non-technical users, managers, and policymakers.
Format	Structured and often lengthy, resembling technical documentation.	Compact and visually oriented, resembling an info card or summary report.

DATA BIOGRAPHY

What Is Data Biography?

In her paper, "An Introduction to the Data Biography," Heather Krause states "Building a data biography, a comprehensive background of the conception, birth, and life of any dataset is an essential step along the path to equity in data science. This includes all data: data that you collected, data that comes from large, trusted sources, data that comes from open data libraries, and data that comes from peer-reviewed research. All data."

Krause's short version of the data biography consists of five core questions:

1. Who
   - Who collected the data?
   - Who owns the data?

2. How
   - The methods behind the data collection design and process.

3. Where
   - In which locations was the data collected?
   - Where is the data stored?

4. Why
   - For what purpose was the data collected?

5. When
   - When was the data collected?[9]

In keeping with Krause's version of a data biography and applying it more broadly to the field of AI, a data biography then refers to a detailed narrative or lifecycle documentation that captures the origin, collection process, transformation, curation, and usage history of a dataset used in an AI system.

It includes information about the dataset's sources, the methodologies applied during preprocessing, the evolution of the dataset over time, and any ethical considerations or biases identified during its lifecycle. The aim is to provide comprehensive transparency into the "life" of the data, from its inception to its application in AI models.

Data Biography – Importance for Assessing Internal Controls

For an auditor or risk management professional tasked with assessing an AI system's internal controls or evaluating its risk posture, a data biography provides important insights into:

1. Transparency: It provides clear visibility into the dataset's origins, transformations, and intended use, enabling auditors to verify compliance with ethical, legal, and organizational standards.
2. Bias Identification: Detailing the dataset's sources and preprocessing methods helps auditors and risk assessors to identify potential biases or ethical issues embedded in the data that could affect the AI system's fairness and performance.
3. Traceability: It ensures that all changes to the dataset, including updates or alterations, are documented, which is critical for validating version control and reproducibility.
4. Risk Assessment: Risk management professionals (as well as auditors) can evaluate risks related to data quality, representativeness, and alignment with the AI system's goals, ensuring that the dataset supports robust decision-making.
5. Compliance Verification: A data biography aids in verifying adherence to data governance policies, privacy regulations, and industry standards, which are key to internal controls.

A data biography is a critical component of AI documentation, offering a granular and holistic view of the dataset's lifecycle, which is invaluable for auditors in evaluating the reliability, accountability, and integrity of the AI system.

Auditing an AI Application's Data Biography

The following questions allow the auditor or risk assessor to examine the processes, documentation, and safeguards surrounding the dataset, promoting a thorough understanding of its lifecycle and how it aligns with the organization's internal controls and ethical standards.

Data Source and Collection

1. What are the primary sources of data for this AI system, and how were these sources selected?
2. How was the data collected, and what specific methodologies or tools were used to ensure its accuracy and completeness?
3. What measures have been taken to ensure that the data sources are representative of the population or problem being addressed by the AI system?

Data Quality and Preprocessing

4. What processes are in place to identify and address missing, duplicate, or erroneous data before it is used in the AI model?
5. What transformations or preprocessing steps have been applied to the raw data, and how were these steps documented?
6. How do you evaluate the quality and reliability of the data at various stages of its lifecycle?

Data Governance and Versioning

7. What version control mechanisms are in place to track changes to the dataset over time, and how is this information documented?
8. How are access permissions to the dataset managed, and what safeguards are in place to prevent unauthorized modifications?
9. What protocols are in place to ensure compliance with data governance policies, privacy laws, and industry standards?

Bias and Ethical Considerations

10. How has the dataset been assessed for potential biases, and what steps have been taken to mitigate any identified biases?
11. What ethical considerations were evaluated during the dataset's collection, preprocessing, and curation, and how are these reflected in its documentation?

Dataset Evolution and Maintenance

12. How frequently is the dataset updated, and what criteria determine when updates are necessary?
13. What mechanisms are in place to assess the impact of updates to the dataset on the performance of the AI model?
14. How is feedback from users or stakeholders incorporated into the evolution and maintenance of the dataset?

Transparency and Traceability

15. What steps have been taken to ensure that the dataset's origin, processing, and usage history are fully transparent and traceable for both internal and external stakeholders?

DATA PORTRAITS

What Is a Data Portrait?

A data portrait refers to a visual or structured representation that summarizes and illustrates the characteristics, composition, or patterns within a dataset.

Marone and Van Durme present the concept of data portraits as AI artifacts that record training data and allow for efficient inspection. A critical property of a data portrait is membership inference: whether an example was part of a data collection.

Furthering the concept of data portraits, Marone and Van Durme, considered three perspectives of the data:

1. Content creators (owners), scientists, and content consumers. Creators wish to know whether a dataset contains their content (e.g., copywritten code).
2. Scientists may seek to understand model behavior by assessing test set leakage or knowledge memorization.
3. Content consumers (including downstream applications) may want to know if a model plagiarizes or cites existing resources.

Datasets should be open and transparent. However, in cases where this is impossible, the authors maintain that certain implementations of data portraits can still allow for (approximate) membership testing on the dataset used to build a specific model.[10]

The concept of an AI artifact, data portraits, as described, is a relatively new idea, approach, concept, and addition to the growing list of AI documentation, useful in sustaining and substantiating AI system openness and transparency.

Data portraits, when developed as proposed by Marone and Van Durme have as their objective, to provide insights into the dataset's attributes, such as distributions, relationships, biases, or trends, by offering a snapshot of its content in a way that is interpretable by both technical and non-technical stakeholders. These portraits can serve as tools for exploring, understanding, and auditing datasets used in AI systems.

This concept of AI "data portrait" artifacts and what insight these artifacts can provide into the integrity of machine language training datasets and the data's membership in those datasets has practical value to the auditor and risk assessor when examining an AI sytem/application's internal controls and risk posture.

The initial question that the auditor or risk assessor should ask the AI development team and AI application users, is "Are data portraits utilized in the AI development/acquisition/implementation process as one tool to gauge/assess openness, transparency, and overall governance in the organization's AI applications?"

The Audit Value of Data Portraits

Data portraits, in their role as an AI documentation tool, may not be in use within the reader's organization now, but they may ultimately be. Based on further examination (risk assessment), recommending the incorporation and use of data portraits may be included in the auditor's final report to management.

Data portraits can provide significant value to an IT auditor or risk assessor examining the internal controls or risks inherent within an AI system/application, for the following reasons:

- Transparency into Dataset Composition: Data portraits crafted correctly are designed to offer a visual or structured representation of datasets, helping risk assessment professionals (auditors and risk management professionals) understand the data's composition, distribution, and key characteristics. This allows for the identification of potential anomalies, imbalances, or biases in the dataset that could compromise the AI system's reliability or fairness.
- Identification of Data Biases: By visualizing the demographic, categorical, or statistical attributes of the dataset through a data portrait, auditors can evaluate whether the data introduces inherent biases that could affect the AI system's outputs or decision-making processes.
- Insight into Data Provenance and Lifecycle: Data portraits should include metadata about the origin, transformations, and lineage of the data, supporting openness and transparency. This information helps auditors verify compliance with data governance policies, ethical guidelines, and legal requirements.
- Evaluation of Data Integrity and Completeness: Through data portraits, auditors can assess whether the dataset is complete, consistent, and well-documented. This ensures the AI system operates on accurate and well-maintained data, which is crucial for maintaining control reliability.
- Support for Risk Assessment: By providing a high-level view of data quality and trends, data portraits can highlight potential risks, such as overfitting, under-representation of critical subgroups, or non-compliance with privacy regulations like GDPR or CCPA.
- Facilitating Stakeholder Communication: Data portraits translate complex datasets into digestible visuals or summaries, aiding auditors in explaining their findings to stakeholders who may not have technical expertise, such as management or regulatory bodies.
- Assessing AI internal controls using the concept of AI data portraits has an overall benefit to the organization (executive management as well as end-users) by enabling IT auditors to assess the underlying dataset's integrity, quality, and governance, which are foundational

to the effectiveness, reliability, and accountability of the overall AI system.

Auditing Data Portrait Artifacts

The following questions may be incorporated into the auditor or risk assessor's examination of an AI system/application if the organization utilizes data portraits as an AI documentation tool. These questions are designed to assist in the examination of how data portraits contribute to accountability, transparency, and risk management within AI systems.

The reader may disregard these questions if data portraits are not currently used within the organization by the AI team or required by users. The auditor should, however, raise the question of why data portraits are not utilized by the AI team or required by end-users.

Data Provenance and Membership

1. How do the data portraits document the provenance of the dataset, including its sources, transformations, and ownership information?
2. What mechanisms within the data portraits allow for the verification of whether specific data instances were included in the training dataset?

Bias and Representation

3. How does the data portrait highlight potential biases or imbalances in the dataset, and what methods are used to detect and address these issues?
4. What insights does the data portrait provide about the representativeness of the dataset with respect to the AI system's intended use cases?

Transparency and Governance

5. How does the data portrait enhance transparency regarding the data used in training the AI model, particularly for external stakeholders or auditors?
6. What specific controls or processes does the data portrait reveal about compliance with data governance standards and privacy regulations?

Risk Identification and Mitigation

7. What risks associated with data leakage, test set contamination, or IP violations are identified through the data portrait?
8. How does the data portrait support the identification and mitigation of risks related to ethical concerns, such as the inclusion of sensitive or inappropriate data?

Practical Utility and Performance

9. How does the use of data portraits improve the efficiency of auditing large-scale datasets, and what limitations or trade-offs are involved in their application?
10. In what ways do data portraits support ongoing monitoring and validation of datasets as they evolve over time?

MODEL CARDS

What Are Model Cards?

> AI Model Cards are documentation tools for AI models that provide essential information about the model's intended use, performance, limitations, and ethical considerations. They are designed to improve transparency, accountability, and responsible AI development by offering detailed insights into various aspects of the AI model.

Mitchell et al., writing in "Model Cards For Model Reporting," state that model cards serve to disclose information about a trained ML model. This includes how it was built, what assumptions were made during its development, what type of model behavior different cultural, demographic, or phenotypic population groups may experience, and an evaluation of how well the model performs for those groups.

Model cards are short documents accompanying trained ML models that provide benchmarked evaluation in a variety of conditions, such as across different cultural, demographic, or phenotypic groups (e.g., race, geographic location, sex, Fitzpatrick skin type) and intersectional groups (e.g., age and race, or sex and Fitzpatrick skin type) that are relevant to the intended application domains. Model cards also disclose the context in which models are intended to be used, details of the performance evaluation procedures, and other relevant information.

Each model card could be accompanied by datasheets, nutrition labels, data statements, or factsheets describing datasets that the model was trained and evaluated on. Model cards provide a way to inform users about what ML systems can and cannot do, the types of errors they make, and additional steps that could create more fair and inclusive outcomes with the technology.[11]

Model cards, as proposed by Mitchell et al., contain a set of sections with details that can provide stakeholders with disclosed information about a trained ML model.

Mitchell et al.'s model card, as shown in Figure 16.1, is intended to provide relevant details to consider but is not intended to be complete or exhaustive and may be tailored depending on the model, context, and stakeholders.

The reader unfamiliar with AI Model Cards is directed to the following websites for additional information and further study:

For an API model:
AI Model Card – Learning Resource #1
[https://developers.perspectiveapi.com/s/about-the-api-model-cards?language=en_US]
For a lightweight model to predict 3D facial surface geometry:
AI Model Card – Learning Resource #2
[https://drive.google.com/file/d/1QvwWNfFoweGVjsXF3DXzcrCnz-mx-Lha/preview]

Model Card

- **Model Details**. Basic information about the model.
  - Person or organization developing model
  - Model date
  - Model version
  - Model type
  - Information about training algorithms, parameters, fairness constraints or other applied approaches, and features
  - Paper or other resource for more information
  - Citation details
  - License
  - Where to send questions or comments about the model
- **Intended Use**. Use cases that were envisioned during development.
  - Primary intended uses
  - Primary intended users
  - Out-of-scope use cases
- **Factors**. Factors could include demographic or phenotypic groups, environmental conditions, technical attributes, etc.
  - Relevant factors
  - Evaluation factors
- **Metrics**. Metrics should be chosen to reflect the potential real-world impacts of the model.
  - Model performance measures
  - Decision thresholds
  - Variation approaches
- **Evaluation Data**. Details on the dataset(s) used for the quantitative analyses in the card.
  - Datasets
  - Motivation
  - Preprocessing
- **Training Data**. May not be possible to provide in practice. When possible, this section should mirror Evaluation Data. If such detail is not possible, minimal allowable information should be provided here, such as details of the distribution over various factors in the training datasets.
- **Quantitative Analyses**
  - Unitary results
  - Intersectional results
- **Ethical Considerations**
- **Caveats and Recommendations**

Figure 16.1 Summary of Model Card Sections and Suggested Prompts for Each.[12]

Key Components of AI Model Cards

1. Model Details: Information about the model architecture, training data, and version.
2. Intended Use: Specific applications and environments for which the model is designed.
3. Performance Metrics: Quantitative evaluations of the model's accuracy, precision, recall, and other relevant metrics.
4. Ethical Considerations: Discussion of potential biases, fairness, and ethical implications.
5. Limitations: Known limitations of the model, including scenarios where the model may not perform well.
6. Maintenance: Guidelines for model updates and retraining to ensure continued performance and relevance.

Model Cards: Usefulness in Auditing AI Tools

Using AI Model Cards during an audit can significantly enhance the evaluation process by providing a structured and comprehensive overview of the AI tool. They are most appropriately used to assess areas related to:

- Transparency: Understanding how the model works and its underlying assumptions.
- Performance: Verifying the model's accuracy and robustness across different scenarios.
- Fairness and Bias: Identifying and addressing potential biases and ensuring the model's fairness.
- Compliance and Ethical Use: Ensuring the model adheres to relevant regulations and ethical standards.

Auditing the AI Tool Using the Tool's AI Model Card

- Model Transparency
  1. What are the key architectural components of the AI model?
  2. How was the training data selected, and what preprocessing steps were applied?

- Performance Evaluation
  1. What are the primary performance metrics used to evaluate the model?
  2. How does the model perform across different demographic groups or use cases?

- Intended Use
  1. What are the specific applications and environments for which the model is designed?

2. Are there any scenarios where the model's performance is known to degrade?

- Ethical Considerations
  1. What potential biases have been identified in the model, and what steps have been taken to mitigate them?
  2. How does the model ensure fairness and avoid discriminatory outcomes?

- Limitations and Risks
  1. What are the known limitations of the model?
  2. What risks are associated with the use of this model, and how are they being managed?

- Maintenance and Updates
  1. What are the guidelines for updating and retraining the model?
  2. How often is the model reviewed and updated to ensure it remains current and effective?

- Compliance
  1. How does the model comply with relevant regulations and standards?
  2. Are there documented procedures for handling data privacy and security?

- User and Stakeholder Feedback
  1. How is feedback from users and stakeholders collected and integrated into the model development process?

- Ethical and Responsible AI Use
  1. What ethical guidelines were followed during the development of the model?
  2. How does the organization ensure the responsible use of the AI model?

- Transparency and Accountability
  1. How are the decisions made by the AI model documented and explained?
  2. What mechanisms are in place to ensure accountability for the model's performance and impacts?

Model Development and Validation

1. How does your organization select appropriate models for AI development, and what criteria are used to evaluate their suitability?
2. Can you describe the process for training AI models, including the selection of training data and the handling of potential biases?

3. What procedures are in place to validate AI models before they are deployed, and how is the validation process documented?
4. How does your organization ensure that the AI model's performance is rigorously tested under various conditions before deployment?
5. What mechanisms are used to continuously monitor the performance of AI models after they have been deployed?
6. How does your organization address and mitigate potential biases that may emerge during the model development and validation processes?
7. What steps are taken to ensure the fairness of AI models across different demographic groups or other relevant categories?
8. How is the retraining or updating of AI models managed to ensure ongoing accuracy and relevance?
9. Can you provide examples of how model performance issues have been identified and resolved through monitoring or validation processes?
10. What documentation is maintained to ensure transparency in the model development, validation, and monitoring processes, and how is this information made accessible to relevant stakeholders?

Model Performance

Ground Truth Dataset

A ground truth dataset refers to a collection of data that is accurately labeled and used as a reference to train, test, and validate machine learning (ML) models. It serves as the benchmark against which the model's predictions are compared to determine accuracy. Ground truth data is considered the "gold standard" because it is assumed to be correct and free of errors, providing a reliable basis for assessing the performance of AI models.

Role of Ground Truth Dataset in AI Models

- Data Integrity: Ground truth datasets are essential for ensuring the integrity of the data used in AI models. Accurate labeling and high-quality data prevent the introduction of errors and biases in the model training process.
- Accuracy: By providing a reference point, ground truth datasets enable the measurement of an AI model's accuracy. The model's predictions are compared against the ground truth to calculate performance metrics like precision, recall, and F1 score (for further information on calculating an AI model's F1 score, see Appendix 4).
- Mitigating Bias: Ground truth datasets help identify and mitigate biases in AI models. If the dataset is diverse and representative, it can reduce the likelihood of the model developing biased outcomes.

- Testing and Validation: Ground truth datasets are used during the testing and validation phases to ensure the model generalizes well to new, unseen data. They provide a reliable means to evaluate the model's performance before deployment.

Audit Questions – AI Model Performance

1. Data Collection and Labeling: How was the ground truth dataset collected and labeled? Were there any challenges faced during this process?
2. Dataset Diversity: Does the ground truth dataset include diverse and representative samples relevant to the model's intended application? How is this substantiated?
3. Bias Assessment: What documentation exists to verify steps were taken to identify and mitigate bias in the ground truth dataset?
4. Data Quality Assurance: How is the quality of the ground truth dataset ensured? Are there procedures in place for regular data quality checks?
5. Relevance to Use Case: How closely does the ground truth dataset represent the real-world scenarios the AI model will be applied to?
6. Documentation and Transparency: Is there comprehensive documentation detailing the data sources, labeling criteria, and potential limitations of the ground truth dataset?
7. Validation Procedures: What methods were used to validate the accuracy and reliability of the ground truth dataset? How were these methods and outcomes documented? How were corrective actions (if required) taken, implemented, and documented?
8. Update Frequency: How frequently is the ground truth dataset updated to reflect new information or changing conditions?
9. Model Performance Metrics: What performance metrics were used to evaluate the model against the ground truth dataset, and how were these metrics chosen?
10. Independent Review: Has the ground truth dataset been reviewed or audited by an independent third party to verify its accuracy and completeness? If so, when? What were the results of that audit? Were the results shared with executive management? Stakeholders?

NUTRITION LABELS

What Is a Model Nutrition Label?

- Definition: Simplified representations of an AI model's performance, limitations, and characteristics in a format that resembles nutrition labels. It communicates key metrics like accuracy, bias levels, fairness, and the model's capabilities, limitations, and safety considerations, much like a nutrition label would for food.

- Use in Audit: Helps auditors assess transparency and usability by examining the model's strengths and weaknesses in an easily understandable manner.
- See Appendix 7 – AI Model Nutrition Label.

Audit Questions – Model Nutrition Labels

1. Does the label include disclaimers about the model's intended uses and limitations? How is this disclaimer documented?
2. Does the nutrition label provide an overview of the AI model's accuracy and error rates? How is this information substantiated? How is this documented?
3. How accessible and understandable is the nutrition label to non-technical stakeholders?
4. How are third-party verifications or validations for the information on the label documented?
5. How does the nutrition label represent potential biases in the model's predictions?
6. How transparent is the label about datasets used for model training and evaluation?
7. What process exists to update the label regularly to reflect changes in the model or dataset?
8. What evidence exists that fairness metrics are included in the label, and do they comply with regulatory requirements?
9. What measures have been taken to ensure the accuracy of the information on the label?
10. What metrics are provided on the label and are they adequate for understanding the model's performance in critical scenarios?

DIRECTED ACYCLIC GRAPH (DAG) CARDS

What Is a DAG Card?

As AI applications and the ML that drives them continue to grow and dominate software markets, an understanding of how these applications work and reach decisions is still in its infancy. ML models are harder to test, and debug compared with traditional software.

As discussed previously, model cards were introduced as one form of AI documentation designed to accompany AI models and to provide clear, standardized documentation about an AI model's characteristics, intended use, limitations, and ethical considerations. In the rapidly evolving world of AI systems, model cards alone may be insufficient in assisting developers, users, and auditors in understanding the model's behavior, capabilities, and risks involving the ML behind the model's behavior.

Enter DAG cards.

In the field of AI, a DAG card is an AI documentation artifact designed to describe the structure and flow of data and computations within an AI pipeline or system, particularly when represented as a DAG.[13]

What exactly is a DAG? Broken into its parts a DAG is defined as follows:

Directed: Indicates that the connections (edges) between nodes have a direction, showing a one-way flow or relationship.

Acyclic: Specifies that there are no loops or cycles, meaning you cannot start at one node and follow the edges to return to the same node.

Graph: Refers to a mathematical structure consisting of nodes (vertices) and connections (edges) used to represent relationships or processes. See Figure 16.2.

When used as an AI documentation tool, a DAG card provides details on the sequence of processes, dependencies, and transformations applied to data as it moves through the system, enabling a better understanding and traceability of workflows.

DAG cards are not yet as widely recognized or standardized as other documentation artifacts like datasheets or model cards.

However, they are becoming increasingly relevant in complex AI systems where understanding data lineage, processing pipelines, and dependency management is critical, such as in ML pipelines implemented with tools like Apache Airflow (an open-source platform for authoring, scheduling, and monitoring data and computing workflows), Kubeflow (an open-source platform designed to simplify the deployment, orchestration, and management of ML workflows on Kubernetes), or TensorFlow Extended (an end-to-end open-source platform for deploying production-grade ML pipelines).

How Are DAG Cards Used?

DAG cards will find increasing usage in areas such as:

- Workflow Visualization: DAG cards document the structure of an AI pipeline, highlighting key steps such as data preprocessing, feature engineering, model training, evaluation, and deployment.
- Traceability and Debugging: They enable AI design and development teams (including audit and risk management) to trace errors or inefficiencies within the pipeline by providing clear documentation of each step and its dependencies.
- Auditing and Compliance: DAG cards support auditors in understanding how data flows through the system, ensuring compliance

with data governance policies, and identifying risks like data leakage or unintentional biases.

- Collaboration and Maintenance: By providing a shared understanding of the pipeline's architecture, DAG cards facilitate collaboration among multidisciplinary teams and simplify the onboarding of new team members.

Auditing DAG Card Documentation Artifacts

While their adoption is growing in organizations with sophisticated AI workflows, standardization of DAG Cards as a documentation artifact in AI is still in progress.

The following questions may be incorporated into the auditor's or risk assessor's examination of an AI system/application if the organization utilizes DAG cards as an AI documentation tool. These questions are designed to assess the completeness, accuracy, and utility of DAG cards as documentation artifacts in auditing AI systems.

The reader may disregard these questions if DAG cards are not yet utilized within the organization by the AI team or required by users. The auditor should, however, raise the question of why DAG cards are not utilized by the AI team or required by end-users.

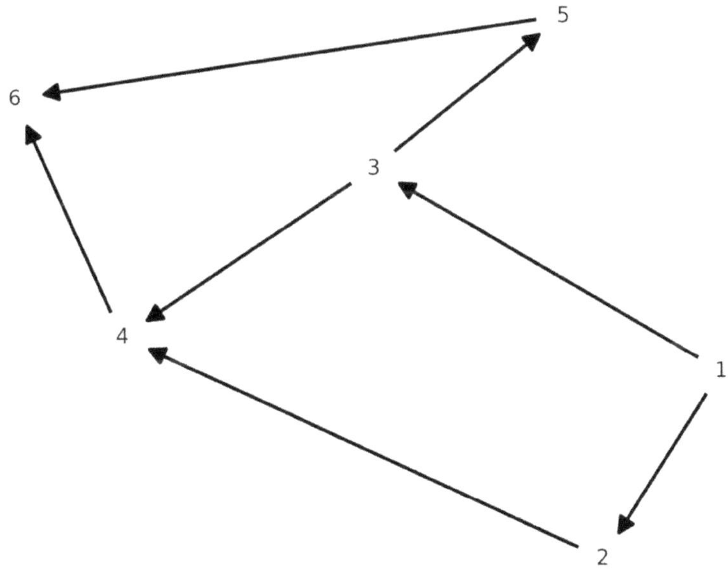

Figure 16.2 Example Of A Directed Acyclic Graph (DAG).[14]

Pipeline Structure and Documentation

1. How does the DAG card represent the structure and dependencies of the AI pipeline, and what insights does it provide about the data flow and task sequencing?
2. What information does the DAG card provide about the parameters, input files, and configurations used in the pipeline, and how is this information tracked for reproducibility?

Versioning and Traceability

3. How does the DAG card support versioning of pipeline artifacts, such as datasets, models, and code, and what processes are in place to ensure traceability across pipeline runs?
4. What mechanisms are documented in the DAG card for tracking changes to pipeline components over time, and how are these used to ensure compliance with internal controls?

Behavioral and Performance Testing

5. What types of behavioral tests are included in the DAG card, and how do they help identify potential performance deficits, edge cases, or biases in the AI system?
6. How does the DAG card document performance metrics, such as training and validation losses, and how are these metrics evaluated across different runs of the pipeline?

Ownership and Accountability

7. What does the DAG card reveal about the ownership and involvement of team members in different pipeline components, and how does this align with organizational accountability policies?
8. How does the DAG card document collaboration between technical and non-technical stakeholders, such as engineering and product teams, in pipeline design and testing?

Efficiency and Transparency

9. How does the DAG card enhance transparency in pipeline operations, particularly regarding resource allocation (e.g., GPU usage) and task prioritization?
10. What measures are documented in the DAG card to optimize the efficiency and reliability of the pipeline, such as mechanisms for handling task failures or parallel execution?

DATA STATEMENTS

What Is a Model Data Statement?

- Definition: Formalized descriptions of the linguistic data or datasets used to train AI systems, including details about origins, demographics, and collection methods. Model data statement ensures clarity and transparency about the data that powers the AI model highlighting its sources, processing, and ethical considerations while addressing potential limitations.

- Use in Audit: Ensures that datasets were created and sourced ethically while meeting diversity and representation standards.
- See Appendix 8 – AI Model Data Statement.

Auditing – Model Data Statements

1. Are there examples provided of appropriate and inappropriate uses of the dataset? How are these examples documented? Retained?
2. By what means does the data statement identify all stakeholders involved in the dataset creation?
3. How are data privacy and consent considerations detailed in the statement?
4. How are disclosures about any known limitations or biases in the dataset documented?
5. How comprehensive is the documentation of the dataset's collection process?
6. How does the data statement address potential gaps in representation or inclusion?
7. How does the statement ensure compliance with international data standards (e.g., GDPR)?
8. How does the statement identify and clearly define the intended audience of the data?
9. What details about the dataset's origins and demographics are provided in the data statement?
10. What evidence is retained/documented in the statement to indicate whether third-party datasets were used, and how they were validated?

FACT SHEETS

What Is a Model Fact Sheet?

A fact sheet is a collection of relevant information (facts) about creating and deploying an AI model or service. Facts could range from information about the purpose and criticality of the model, measured characteristics of the dataset, model, or service, or actions taken during the creation and deployment process of the model or service.

Such models are created by various roles in the ML lifecycle, the collection of steps used to develop and deploy the ML model, such as a business owner, a data scientist, or a person who collects and transforms datasets and ML algorithms to construct the ML model, a model validator, or a person who tests, or validates, the AI model developed by a data scientist, and model operator, or a person who deploys and monitors the AI model so that others can call it.

Each of the many roles in the lifecycle contributes facts about how the model was created and deployed. For example, the business owner can provide the intended use for the model. The data scientist can describe various data-gathering and data-manipulation activities. A model tester can describe key testing measurements, and the model operator can provide key performance metrics. Together, these facts provide a rich story about the construction of the model, similar to how a school transcript or resume can offer more insight into a student or a job applicant.

A fact sheet is modeled after a supplier's declaration of conformity, used in many industries to show a product conforms to a standard or technical regulation. However, fact sheets can be rendered in many different formats, not just printed documents. Fact sheets vary based on the purpose of communication and the audience.[15]

Summary

- Definition: High-level documentation that provides comprehensive information about an AI system, including its purpose, intended uses, limitations, and governance policies.
- Use in Audit: Assists auditors in verifying compliance, understanding risk, and evaluating operational transparency.
- See Appendix 9 – AI Model Factsheet – Model GPT-4 Example.

Audit Questions – Model Fact Sheets

1. How are maintenance and update protocols specified in the fact sheet?
2. How are risk assessment and management strategies detailed in the fact sheet?
3. How does the fact sheet address stakeholder feedback and complaint resolution processes?
4. How does the fact sheet describe the system's compliance with relevant regulations and standards?
5. How does the fact sheet explain the governance framework for the AI tool/system?
6. How transparent is the documentation on the datasets and algorithms used?
7. What documentation exists to substantiate that ethical considerations and responsible AI principles are addressed in the fact sheet?
8. What evidence exists that the fact sheet clearly outlines the AI system's purpose and intended applications?
9. What limitations of the AI system are highlighted in the fact sheet, and how are they mitigated?
10. What performance benchmarks and metrics are provided for the AI system?

SYSTEM CARDS

What Are System Cards?

> AI System Cards are comprehensive documents that provide detailed information about an entire AI system, including its components, operational context, performance, and ethical considerations. An AI system card is a summary matrix score of various key parameters for rapidly changing version control.

A system card in AI is a document or resource that provides a detailed overview of how an AI system operates, its capabilities, limitations, and other relevant aspects. The primary purpose of a system card is to promote transparency, accountability, and responsible use of AI technologies by offering stakeholders – such as users, developers, or regulators – key information about the system.

There are four sections in every AI system card:

1. An overview of the AI system.
2. A section explaining how the AI system works, which includes a summary of the steps involved in creating experiences on Facebook and Instagram.
3. A section describing how to customize the content that is shown. This includes descriptions of system controls and instructions for how each person can control and customize their experience.
4. A section explaining how the AI delivers content, which includes an explanation of how some of the significant prediction models inform the overall AI system and produce product experiences.[16]

Key Components of a System Card

Overview of the System

- Description of the AI model or system.
- Intended purpose and use cases.
- Deployment context and target audience.
- Main system components.

Operational Context

- Information about the environment in which the AI system operates.
- Include data sources and integration points.

Capabilities

- What the system can do effectively.
- Benchmarks, metrics, or evaluations demonstrating its performance.

Limitations
- Known weaknesses or areas where the system performs poorly.
- Risks of misuse, biases, or unintended consequences.

Ethical Considerations
- Potential societal impacts (e.g., on privacy, equity, or inclusion).
- Mechanisms to mitigate harm, such as content moderation or user guidelines.
- Analysis of the potential biases of the AI system.

Safety and Security
- Safeguards in place to ensure reliability and robustness.
- Steps taken to protect user data and prevent adversarial attacks.

Risk Management
- Identification and mitigation of risks associated with the AI system.

Training Data
- Overview of datasets used to train the model.
- Information on data diversity and potential biases.

Evaluation
- How the system was tested (e.g., real-world simulations, lab environments).
- Results of tests across different demographics or scenarios.

Performance Metrics
- Detailed performance evaluation, including accuracy, reliability, and efficiency metrics across various scenarios.

Content Delivery
- How the AI delivers content, which includes an explanation of how some of the significant prediction models inform the overall AI system and produce product experiences.

Updates and Iterations
- How the system evolves over time.
- Version control or changelogs.
- Procedures for system updates, monitoring, and continuous improve ment.

User Guidance
- Instructions on appropriate and inappropriate use.
- FAQs or troubleshooting support.

System cards are similar in concept to model cards or datasheets for datasets, all of which aim to standardize how information about AI systems is shared. See Appendix 10 – AI Model System Card.

Why Are System Cards Important?

Systems cards are important in three specific ways:

1. Trust Building: They provide clarity, enabling users to understand the system's design and constraints.
2. Accountability: Developers can be held responsible for the system's outcomes based on disclosed details.
3. Compliance: They align with ethical AI principles and legal regulations by offering transparency.

The reader seeking additional information on AI system cards is directed to the following websites for further study:

What Is a Sy]stem Card?
AI System Card – Learning Resource #1
[https://ai.stackexchange.com/questions/42670/what-is-system-card]
Meta's System Cards
AI System Card – Learning Resource #2
[https://siliconangle.com/2023/06/29/metas-new-system-cards-explain-ai-powers-content-recommendations-facebook-instagram/]
GPT-4 System Card
AI System Card – Learning Resource #3
[https://cdn.openai.com/papers/gpt-4-system-card.pdf]

Using the System Card to Audit AI Tools

Using AI system cards during an audit can significantly enhance the evaluation process by providing a structured and comprehensive overview of the entire AI system.

They are most appropriately used to assess areas related to:

- Transparency: Understanding the complete architecture and operational context of the AI system.
- Performance: Verifying the system's reliability and efficiency across different scenarios and environments.
- Ethical and Social Impact: Identifying and addressing ethical issues and potential biases.
- Risk Management: Ensuring the system's risks are identified and mitigated.

Auditing AI System Cards

- System Architecture
  1. What are the main components of the AI system, and how do they interact?
  2. What data sources are integrated into the AI system, and how is data quality ensured?
- Operational Context
  1. In what environments and applications is the AI system deployed?
  2. How does the system handle different operational scenarios and edge cases?
- Performance Evaluation
  1. What are the key performance metrics used to evaluate the system?
  2. How does the system perform under varying loads and conditions?
- Ethical and Social Impact
  1. What ethical constructs have been evaluated during the development of the system?
  2. How does the system ensure fairness and avoid biases in its outputs?
- Risk Management
  1. What are the primary risks associated with the AI system, and how are they mitigated?
  2. How is the system monitored for potential failures or adverse outcomes?
- Maintenance and Updates
  1. What procedures are in place for updating and maintaining the AI system?
  2. How frequently is the system reviewed and updated to ensure ongoing effectiveness?
- Compliance
  1. How does the AI system comply with relevant regulations and standards?
  2. Are there documented procedures for handling data privacy and security within the system?
- User and Stakeholder Feedback
  1. How is feedback from users and stakeholders collected and incorporated into system improvements?
- Transparency and Accountability
  1. How are the decisions made by the AI system documented and explained?
  2. What mechanisms are in place to ensure accountability for the system's performance and impacts?

- Ethical and Responsible AI Use
  1. What ethical guidelines were followed during the development and deployment of the system?
  2. How does the organization ensure the responsible use of the AI system?

METHOD CARDS

What Are Method Cards?

> The use of Method Cards is an emerging practice and is less prevalent compared to established documentation artifacts like Model Cards and Datasheets. However, their adoption is gaining momentum, particularly in organizations emphasizing data-centric AI and reproducibility.

Adkins et al. describe techniques such as datasheets, factsheets, and model cards as having taken a mainly descriptive approach, providing various details about the system components.

The authors continue to stress that while the above information is essential for product developers and external experts to assess whether the ML system meets their requirements, other stakeholders might find it less actionable.

In particular, ML engineers need guidance on how to mitigate potential shortcomings to fix bugs or improve the system's performance. Adkins et al. propose a preliminary approach, which the researchers refer to as method cards.[17]

While an emerging artifact of AI documentation, currently available information describes method cards as prescriptive documentation artifacts designed to provide clear guidance on the appropriate use, implementation, and potential limitations of specific ML methods and techniques.

Note to Reader

Prescriptive documentation refers to documentation that provides explicit instructions, recommendations, or guidelines on how to perform tasks, implement processes, or address specific challenges. Unlike descriptive documentation, which simply records and explains existing systems or processes, prescriptive documentation focuses on guiding actions to achieve specific outcomes, resolve issues, or optimize performance.

In the context of AI, prescriptive documentation (e.g., method cards) helps ML engineers to:

- Understand best practices for applying ML methods.
- Make informed decisions about preprocessing, model design, and optimization.
- Address potential issues like biases, failure modes, and inefficiencies.
- Ensure responsible and reproducible use of algorithms and methods.

This type of documentation is action-oriented and particularly valuable for operationalizing systems, troubleshooting, and improving processes. Auditors, along with risk assessors, may find value in examining AI method cards in their process of evaluating the AI system/application.

The primary objective of method cards therefore is to enhance transparency and reproducibility by combining descriptive details with actionable recommendations for various stages of model development, such as data preparation, modeling, optimization, testing, and debugging. In essence, then, method cards serve as a reference for ML engineers and external reviewers, ensuring the responsible and effective application of ML methods.

Method Cards – Value and Use

Method cards, when used by organizations as another level of AI system/application documentation, are particularly valuable for:

- **Internal Controls:** By documenting best practices, common pitfalls, and failure modes, they help ensure robust development processes and reduce operational risks.
- **Risk Management:** They enable auditors, risk assessors, and engineers to evaluate potential biases, ethical concerns, and vulnerabilities within the ML methods used in AI applications.
- **Onboarding and Knowledge Sharing:** They provide standardized, prescriptive guidance that helps onboard new team members and promotes consistent application of ML techniques.

As an emerging tool, method cards are not yet standardized across the industry, but they hold significant potential for improving documentation quality and ensuring the alignment of AI systems with organizational and regulatory requirements.

Auditing Method Card Documentation

The following questions may be incorporated into the auditor's or risk assessor's examination of an AI system/application if the organization utilizes method cards as an AI documentation tool.

These questions aim to evaluate how well method cards contribute to transparency, accountability, and risk management, ensuring that the AI system adheres to sound internal controls and ethical standards.

The reader may disregard these questions if method cards are not yet utilized within the organization by the AI team or required by users. The auditor should, however, raise the question of why method cards are not utilized by the AI team or required by end-users.

General Documentation and Purpose

1. What specific ML methods and techniques are documented in the method cards, and how are these aligned with the AI system's goals and operational requirements?
2. How does the method card articulate the appropriate use cases and limitations of the method, and what safeguards are in place to prevent misuse?

Transparency and Traceability

3. What level of detail does the method card provide about data preparation, modeling, and optimization steps, and how does this enhance traceability across the AI lifecycle?
4. How does the method card support the identification and documentation of failure modes, biases, or privacy concerns related to the ML method?

Reproducibility and Benchmarking

5. How do the method cards ensure that the documented techniques and processes are reproducible, and what evidence exists to validate reproducibility claims?
6. What benchmarking metrics and thresholds are documented in the method card, and how are they used to assess the performance and fairness of the ML method?

Risk Management and Controls

7. How are the risks associated with out-of-distribution behavior, adversarial vulnerabilities, or data/model drift addressed in the method card?
8. What specific guidance does the method card provide on mitigating risks such as biases, false positives, or overfitting during development and deployment?

Practical Utility and Evolution

9. How are method cards updated to reflect new insights, techniques, or operational challenges, and how is this process integrated into the organization's AI governance framework?
10. In what ways do the method cards improve collaboration, onboarding, and knowledge transfer for stakeholders involved in the AI system's lifecycle?

NOTES

1 (n.a.), 2024, "Software System Documentation Definition," Law Insider, www.lawinsider.com/dictionary/software-system-documentation?utm_source=chatgpt.com

2 Metha, S., Rogers, A., and Gilbert, T., March 20, 2023, "Dynamic Documentation for AI Systems," *Association for Computing Machinery*, vol. 1, no. 1, Article, https://arxiv.org/pdf/2303.10854.
3 Dylan and Jess, (n.d.), "The Limitations of ChatGPT with Emily M. Bender and Casey Fiesler," The Radical AI Podcast, podcast marker, 00:39:21, https://sonix.ai/r/iLD3Xv5UriGuArS6kfdtbyv8/transcript.pdf?language=undefined
4 Cuomo, J., September 4, 2023, "The Landscape of Transparent Documentation Standards for AI," Medium, https://medium.com/@JerryCuomo/the-landscape-of-transparent-documentation-standards-for-ai-9b3b434d7fff.
5 In AI, both "data sheets" and "datasheets" are used, but the preferred term is typically "datasheets" (written as one word) when referring to standardized documentation. This convention follows the naming introduced by the influential paper "Datasheets for Datasets" by Timnit Gebru and others, which advocates for structured documentation of datasets in AI. The naming convention "datasheet(s)" is used when specifically referring to or referencing AI system/application documentation.
6 Gebru T., et al., December 1, 2021, Datasheets for Datasets," *Communi cations of the ACM*, vol. 64, no. 12, pp. 86–92, https://dl.acm.org/doi/pdf/10.1145/3458723.
7 Pushkarna, M., Zaldivar, A., and Kjartansson, O., June 20, 2022, "Data Cards: Purposeful and Transparent Dataset Documentation for Responsible AI," In *FAccT '22: Proceedings of the 2022 ACM Conference on Fairness, Accountability, and Transparency*, 1776–1826 Pages, https://dl.acm.org/doi/fullHtml/10.1145/3531146.3533231.
8 Pushkarna, M., and Zaldivar, A., November 17, 2022, "The Data Cards Playbook: A Toolkit for Transparency in Dataset Documentation," Google Research, https://blog.research.google/2022/11/the-data-cards-playbook-toolkit-for.html.
9 Krause, H., January 21, 2019, "An Introduction to the Data Biography," We All Count, https://weallcount.com/2019/01/21/an-introduction-to-the-data-biography.
10 Marone, M., and Van Durme, B., December 14, 2023, "Data Portraits: Recording Foundation Model Training Data," Thirty-seventh Conference on Neural Information Processing Systems Datasets and Benchmarks Track, https://arxiv.org/abs/2303.03919.
11 Margaret Mitchell, Simone Wu, Andrew Zaldivar, Parker Barnes, Lucy Vasserman, Ben Hutchinson, Elena Spitzer, Inioluwa Deborah Raji, and Timnit Gebru, January 29–31, 2019, "Model cards for model reporting," In Proceedings of the Conference on Fairness, Accountability, and Transparency. ACM, 220–229, https://doi.org/10.48550/arXiv.1810.03993.
12 Margaret Mitchell, Simone Wu, Andrew Zaldivar, Parker Barnes, Lucy Vasserman, Ben Hutchinson, Elena Spitzer, Inioluwa Deborah Raji, and Timnit Gebru, January 29-31, 2019, Figure 16.1, Summary Of Model Card Sections And Suggested Prompts For Each, "Model cards for model reporting," In Proceedings of the Conference on Fairness, Accountability, and Transparency. ACM, 220–229, https://doi.org/10.48550/arXiv.1810.03993.

13 A directed acyclic graph (DAG) is a collection of points (called nodes) connected by arrows (called edges), where the arrows show the direction of movement or flow, and there are no loops – meaning you cannot start at one node and follow the arrows to get back to the same node. It is like a one-way street map that never circles back to where it started.

14 OpenAI's ChatGPT, December 14, 2024, Figure 16.2, Directed Acyclic Graph (DAG) produced by ChatGPT based on prompt submitted by the author.

15 Hind, M., (July 9, 2020), "Introduction to AI FactSheets," IBM Research, https://research.ibm.com/blog/aifactsheets.

16 Meta AI, June 29, 2023, "Introducing 22 system cards that explain how AI powers experiences on Facebook and Instagram," https://ai.meta.com/blog/how-ai-powers-experiences-facebook-instagram-system-cards.

17 Adkins, D., Alsallakh, B., and Cheema, A., et al., April 27, 2022, "Prescriptive and Descriptive Approaches to Machine-Learning Transparency," ACM CHI Conference on Human Factors in Computing Systems 2022, https://arxiv.org/pdf/2204.13582.

Chapter 17

Reinforcement Learning Models and Reward Reports

REINFORCEMENT LEARNING MODELS

Reinforcement learning (RL) models are a type of machine learning model where an artificial intelligence (AI) agent learns to make decisions by interacting with an environment and receiving feedback in the form of rewards or penalties. RL models aim to maximize cumulative rewards over time by learning the optimal sequence of actions to achieve a specific goal.

An AI agent is a system that perceives its environment, makes decisions, and takes actions to achieve specific goals. It uses algorithms to process inputs (like data or observations) and determines the best actions to maximize success, often based on feedback or rewards. Examples include virtual assistants, self-driving cars, and game-playing programs.

KEY FEATURES OF RL MODELS

- Agent–Environment Interaction

  The AI agent interacts with its environment, takes actions, and observes outcomes.
- Reward-Based Learning

  The model uses a reward function to evaluate the agent's actions.

  Positive rewards reinforce desirable behavior; penalties discourage undesirable actions.
- Trial and Error

  The agent learns through exploration and exploitation, improving its performance through repeated experiences.
- Policy Optimization

  RL models learn a policy: a mapping from states (current situations) to actions (what the agent should do).

DOI: 10.1201/9781003646501-22

HOW RL MODELS DIFFER FROM OTHER AI MODELS

Supervised Learning

Training Data: Supervised learning uses labeled datasets to train models.
Output: The model learns to predict an output (e.g., classification or regression).
Difference: RL does not use explicit labels; it learns through interaction and rewards.

Unsupervised Learning

Training Data: Unsupervised learning identifies patterns in unlabeled data (e.g., clustering).
Difference: RL focuses on learning optimal decision-making through action–reward cycles, not just discovering data structures.

RL's Unique Aspect

RL is designed for **sequential decision-making** in dynamic environments where decisions affect future outcomes.

Examples: Playing games, controlling robots, or managing autonomous systems.

WHY RL IS UNIQUE IN AI

- RL models learn from experience rather than pre-defined examples.
- The reward system allows them to optimize decisions over time, making RL ideal for tasks requiring strategy, adaptability, and long-term planning.
- RL is particularly suited to complex environments where actions have delayed or cumulative consequences, unlike other AI models that focus on static predictions.

In short: RL models differ because they emphasize learning through action, feedback, and adaptation, which makes them powerful for real-time decision-making tasks.

Gilbert et al. write that RL is one of the most promising branches of foundational research in AI. In the long term, RL is considered by many AI theorists to be the most promising path to artificial general intelligence. This places RL practitioners in a position to design systems that have never existed and lack prior documentation in law and policy.[1]

REWARD FUNCTION AND RL

A reward function or mechanism is a core concept in RL, a subset of machine learning. It defines how an AI system evaluates its performance and decides which actions are "good" or "bad" concerning a specific goal. The reward function specifies how rewards (numerical signals) are assigned based on the AI's actions and outcomes within its environment. This mechanism is crucial for training an AI agent to achieve specific goals.

The reward function is designed to reflect the objectives of the system. For example:

- In a self-driving car, a reward might be given for safely reaching a destination.
- In a chatbot, a reward might reflect accurate and helpful responses to user queries.

  The reward function is critical because it directly shapes the AI's learning process. Poorly designed reward functions can cause:

  - Unintended behavior: The AI optimizes for an unintended result (e.g., maximizing a metric at the cost of ethics or safety).
  - Ethical concerns: Rewards may prioritize efficiency over safety or fairness (e.g., It can expose systems to risks, biases, or unintended consequences).
  - Misalignment: The AI's goals conflict with the organization's objectives. In a financial AI, for example, if the reward function only prioritizes profit, it may ignore compliance rules.

These reports document how rewards are defined, measured, and distributed within an AI model to optimize its outputs and align system performance with desired goals. They include quantitative thresholds, qualitative objectives, and insights into how reward outcomes influence the learning process.

In short, the reward function serves as the AI's objective guide, ensuring it learns behaviors that align with human-defined goals and constraints.

Building on the concept of Model Cards, System Cads, and Data Sheets, Gilbert et al. propose the design and use of Reward Reports as living documents for proposed RL deployments that demarcate design choices. This includes which types of feedback have been brought into scope, what metrics have been considered to optimize performance, and why components of the specification (e.g., states, actions, rewards) were deemed appropriate. Reward Reports also outline the conditions for updating the report following ex-post evaluation of system performance, the domain, or any interaction between these.

The authors also argue that Reward Reports are necessary for design decisions to be exhaustively evaluated and recommend that Reward Reports be used as a tool for AI governance beyond RL, enabling authorities to examine what a given system is optimizing for and to evaluate the appropriateness of those terms. Reward Reports will ensure that design choices are auditable by third parties, contestable through litigation, and able to be affirmed by stakeholders.[2]

WHAT ARE REWARD REPORTS?

Reward Reports are defined as documentation artifacts that describe the reward function or mechanism used to guide and evaluate the behavior of an AI system, particularly in RL models.

The use of Reward Reports as proposed by Gilbert et al. is growing but not yet ubiquitous. Industries like autonomous systems, robotics, and financial AI applications often prioritize them due to their reliance on RL. However, many organizations developing simpler AI applications, particularly those using supervised or unsupervised learning, may not create detailed Reward Reports.

Adoption of Reward Reports as AI documentation artifacts increases when companies face:

- High regulatory scrutiny
- Ethical AI demands
- Significant operational risk.

Reward Reports are becoming more valuable as auditors and regulators demand transparency in AI systems, driving their adoption across sectors where AI accountability is critical.

CAPACITY AND USE OF REWARD REPORTS

Reward Reports are found beneficial as AI documentation artifacts when they are used to:

- Document Reward Function Design: Clearly outline how rewards are calculated, which objectives they align with, and what metrics determine success or failure.
- Ensure Goal Alignment: Validate that rewards encourage behaviors consistent with ethical, operational, and business goals.
- Evaluate Bias and Fairness: Highlight any biases in reward functions that may cause unintended consequences.
- Support Audits and Compliance: Provide a structured, transparent explanation of why AI behaves as it does and the rationale for optimizing specific outputs.

REWARD REPORTS: VALUE TO IT AUDITORS AND RISK ASSESSORS

Auditors and risk assessment professionals will find that Reward Reports (when in use by the organization) are an additional source of information providing details about an AI system/application's internal controls and risk posture.

- Internal Controls Evaluation: Reward Reports help auditors assess if the AI's reward mechanism ensures appropriate control over decision-making processes. Misaligned rewards may incentivize undesirable outcomes.
- Risk Identification: They help auditors identify operational risks, such as bias, unintended behavior, or ethical risks caused by flawed reward structures.
- Transparency: Documenting reward functions increases accountability and clarity for complex systems that rely on opaque decision-making algorithms.
- Compliance and Governance: Assessors can confirm that the system aligns with regulations, policies, and ethical AI principles.

Auditing RL Models Using Reward Reports

Although the use of Reward Reports is currently limited, the following audit questions have been included in this text as the use of Reward Reports is increasing in industries like autonomous systems, robotics, and financial applications, where RL plays a significant role.

The growing demand for transparency, ethical AI governance, and regulatory compliance is driving organizations to adopt Reward Reports as tools for improving accountability and risk management. While not yet ubiquitous, their use is gaining traction as auditors, policymakers, and stakeholders recognize the value of documenting reward functions to ensure alignment with organizational and societal goals.

As regulatory scrutiny and ethical AI demands intensify, Reward Reports are expected to become more common across AI systems development and implementation.

The following questions are provided for the auditor and risk assessor to use when tasked with assessing reward functions as part of RL models.

Design Phase

1. How was the reward function designed, and what objectives or metrics does it aim to optimize?

   Rationale: Helps identify alignment between design goals and real-world outcomes.

2. What measures were taken to prevent reward function misspecification or "reward hacking?"
   Rationale: Addresses risks of unintended behavior due to poorly defined rewards.
3. How were ethical and operational risks considered when defining rewards? Were fairness, bias, and safety addressed?
   Rationale: Ensures ethical considerations are integrated into reward design.
4. What stakeholder input was gathered to validate the appropriateness of the reward structure?
   Rationale: Determines the inclusion of diverse perspectives during design.
5. Were there trade-offs considered when optimizing the reward function, and how were they justified?
   Rationale: Reveals conscious design decisions and priorities.

Development Phase

6. How does the reward mechanism guide the RL model's exploration–exploitation trade-off during training?
   Rationale: Ensures the model balances learning and performance.
7. What testing frameworks were applied to validate that the reward function encourages expected behaviors?
   Rationale: Evaluates rigorous testing of reward alignment.
8. How are long-term and short-term rewards balanced in the system? Is there a discounting mechanism in place?
   Rationale: Assesses whether cumulative rewards align with long-term goals.
9. What fallback mechanisms exist in case the RL model optimizes for unintended behaviors during testing?
   Rationale: Identifies controls for mitigating development risks.

Testing Phase

10. Were simulations or real-world scenarios used to test the reward function, and how realistic were the environments?
    Rationale: Evaluates the robustness of reward testing in dynamic conditions.
11. What anomalies or unexpected behaviors were observed during testing, and how were they addressed?
    Rationale: Identifies issues encountered and remedial measures taken.
12. How was the reward function validated for consistency across different testing environments or datasets?
    Rationale: Ensures the reward function generalizes effectively.

Implementation Phase

13. What monitoring processes are in place to track the RL system's performance and ensure alignment with the reward function?
    Rationale: Assesses ongoing oversight mechanisms.
14. How are updates to the reward function documented post-implementation, and under what conditions are changes made?
    Rationale: Ensures transparency in reward adjustments over time.
15. What governance or audit frameworks are applied to evaluate the appropriateness and fairness of the reward outcomes?
    Rationale: Evaluates compliance with ethical and governance standards.

These questions encourage thorough, articulated responses that address the design choices, risk considerations, and governance measures of RL systems and their reward structures.

NOTES

1 Gilbert, T., Lambert, N., Dean, S., and Zick, T., February 2022, "Choices, Risks, and Reward Reports: Charting Public Policy for Reinforcement Learning Systems," UC Berkeley, Center for Long-Term Cybersecurity, Artificial Intelligence Security Initiative (AISI), https://cltc.berkeley.edu/wp-content/uploads/2022/02/Choices_Risks_Reward_Reports.pdf.

2 Gilbert, T., Lambert, N., Dean, S., and Zick, T., February 2022, "Choices, Risks, and Reward Reports: Charting Public Policy for Reinforcement Learning Systems," UC Berkeley, Center for Long-Term Cybersecurity, Artificial Intelligence Security Initiative (AISI), https://cltc.berkeley.edu/wp-content/uploads/2022/02/Choices_Risks_Reward_Reports.pdf.

Chapter 18

Code Analysis

CODE ANALYSIS

The importance of code review as a software quality assurance activity cannot be overstated. It includes a thorough examination by one or more individuals other than the code's author – known as reviewers. This process is not only about identifying errors but also about maintaining the boundaries between various quality assurance techniques such as static analysis and testing.

Artificial intelligence (AI) code analysis in software development is multifaceted, embracing techniques like static and dynamic code analyses, rule-based systems, and cutting-edge approaches, involving natural language processing and large language models. Static code analysis scrutinizes code without executing it, detecting potential vulnerabilities and ensuring compliance with coding standards. In contrast, dynamic code analysis involves testing and evaluating code during execution, identifying runtime issues that static analysis might miss.[1]

ANALYZING AI CODE

The following is a brief overview of code analysis processes that may be applied to analyzing AI software code.

Static Code Analysis

- Examines the code without executing it.
- Tools like PyLint, Flake8, or SonarQube are used to check for syntax errors, unused variables, coding standards, and security vulnerabilities.
- In AI code, static code focuses on verifying the proper use of libraries (e.g., TensorFlow, PyTorch), ensuring modularity in data preprocessing and model definition, and checking hyperparameter configurations.

DOI: 10.1201/9781003646501-23

Dynamic Code Analysis

- Involves running the code and monitoring its behavior during execution.
- Debuggers and profilers (e.g., PyCharm Debugger, cProfile) help track runtime issues such as memory leaks or inefficient loops.
- For AI, this includes analyzing data processing pipelines, verifying the correctness of gradient calculations, and monitoring training for anomalies like vanishing/exploding gradients.

Algorithmic Analysis

- Ensures that the AI algorithms used are logically sound and aligned with the desired objectives.
- Review of loss functions, optimization methods, and regularization techniques to ensure proper model convergence and generalization.

Data and Input Validation

- Evaluates the data pipeline for proper handling of missing values, scaling, and encoding.
- Ensures training and test datasets are balanced and representative.

Model Behavior Analysis

- Examines the performance and fairness of the AI model using metrics, such as accuracy, precision, recall, F1-score, and bias detection methods.
- Validates interpretability techniques like SHAP or LIME to ensure decisions made by the AI are explainable.

Automation with Tools

- Specialized tools like TensorFlow Model Analysis (TFMA), W&B (Weights and Biases), or PyTorch Lightning integrate static and dynamic checks and facilitate model debugging.

Figure 18.1 is a representative, partial sample of a Python program using a neural network to generate a picture of a dog. Developers can use pre-trained models like Generative Adversarial Networks (GANs) or tools like TensorFlow or PyTorch.

This example uses TensorFlow and a pre-trained model such as a GAN to create the image. For simplicity, this code will set up the environment and load a pre-trained GAN model from TensorFlow Hub, which can generate images of dogs.

TensorFlow is a versatile and powerful framework for machine learning and deep learning. It provides a range of tools and resources that enable developers and researchers to build, train, and deploy sophisticated models across various platforms and applications. TensorFlow provides a

```
import tensorflow as tf
import tensorflow_hub as hub
import numpy as np
import PIL.Image

# Check TensorFlow version
print(f"TensorFlow version: {tf.__version__}")

# Load the pre-trained GAN model from TensorFlow Hub
model_url = 'https://tfhub.dev/google/gan/dog2cat/1'
gan_model = hub.load(model_url)
print("Model loaded successfully!")

# Function to generate a random noise vector
def generate_noise_vector(dimensions):
   return np.random.randn(1, dimensions)

# Function to generate a dog image from the GAN model
def generate_dog_image(model, noise_dim):
   noise_vector = generate_noise_vector(noise_dim)
   generated_image = model(noise_vector)
   return generated_image

# Function to save the generated image
def save_image(image_tensor, file_name):
   image = image_tensor.numpy()
   image = (image * 255).astype(np.uint8)
   image = PIL.Image.fromarray(image[0])
   image.save(file_name)
   print(f"Image saved as {file_name}")
```

Figure 18.1 TensorFlow Python Program Code Example.[2]

comprehensive ecosystem of tools, libraries, and community resources that facilitate the development of machine-learning applications.

AI CODE – AUDIT OBJECTIVE AND PROCEDURE

Audit Objective

Effectively evaluate the TensorFlow code's quality, security, and compliance, ensuring it meets the necessary standards and requirements.

Audit Procedure

To effectively audit the TensorFlow code provided, the auditor should follow a systematic procedure to ensure the code's accuracy, reliability, and compliance with best practices.

Audit Procedure Steps

1. Code Review
    - Examine the code for clarity, readability, and adherence to coding standards.
    - Verify that the code includes appropriate comments and documentation.
2. Environment Setup
    - Check the installation of necessary libraries and dependencies.
    - Ensure that the code is compatible with the specified versions of TensorFlow and other libraries.
3. Model Loading
    - Verify the source and integrity of the pre-trained model loaded from TensorFlow Hub.
    - Confirm that the model URL and loading mechanism are secure and up-to-date.
4. Input Data Validation
    - Ensure th0at the noise vector generation function creates appropriate input data for the GAN model.
    - Validate the range and distribution of the generated noise vectors.
5. Model Execution
    - Run the code to generate an image and observe any errors or warnings.
    - Verify that the generated image meets the expected output format and quality.

6. Output Verification
   - Confirm that the generated image is saved correctly and can be displayed.
   - Check the file format, resolution, and content of the saved image.
7. Security and Compliance
   - Review the code for any potential security vulnerabilities.
   - Ensure that the code complies with relevant data privacy and security standards.
8. Performance and Efficiency
   - Evaluate the performance of the code, including execution time and resource utilization.
   - Check for any unnecessary computations or optimizations that can be made.
9. Documentation and Logging
   - Ensure that the code includes adequate logging for debugging and auditing purposes.
   - Verify that the documentation covers all aspects of the code, including setup, execution, and troubleshooting.
10. Maint]enance and Updates
    - Check for any dependencies on external sources that may require updates.
    - Ensure that the code is designed for maintainability and future modifications.

Auditing AI Code – General Audit Questions

While the audit questions that follow are related to the representative partial sample of Python program code provided in Figure 18.1, the audit procedure and the approach to auditing AI code in other applications would be identical.

The questions asked may need modification to relate to and address the specificity of the code under review. All audit responses must be documented and substantiated. Binary "yes," or "no" responses are not acceptable.

- Code Quality
  1. Is the code well-structured, readable, and adheres to coding standards?
  2. Which coding standards are followed by the developer?
  3. Are there sufficient comments and documentation to understand the code's functionality?

- Environment Setup
  1. Are all necessary libraries and dependencies properly installed and specified?
  2. Is the code compatible with the specified versions of TensorFlow and other libraries?
- Model Integrity
  1. Is the pre-trained model sourced from a reliable and secure location?
  2. Are there any checks to ensure the integrity of the model being loaded?
- Input Data
  1. Is the noise vector generation function creating appropriate input data for the GAN model? *Refer to Appendix 3 for additional information on vector generation functions and completing this audit step.*
  2. Are the range and distribution of the generated noise vectors valid for the model?
- Model Execution (this step requires the auditor to request a copy of the code be copied and placed in a test library and the auditor would execute the code from the test library)
  1. Does the code execute without errors or warnings?
  2. Does the generated image meet the expected output format and quality?
- Output Verification
  1. Is the generated image saved correctly in the specified file format?
  2. Is the content of the saved image as expected, and can it be displayed without issues?
- Security
  1. Are there any potential security vulnerabilities in the code, such as dependency on untrusted sources?
  2. Does the code comply with relevant data privacy and security standards?
- Performance
  1. How efficient is the code in terms of execution time and resource utilization?
  2. Are there any unnecessary computations or optimizations that can be made?
- Logging and Documentation
  1. Does the code include adequate logging for debugging and auditing purposes?
  2. Is the documentation comprehensive, covering setup, execution, and troubleshooting?

- Maintenance
  1. Are there any dependencies on external sources that may require updates?
  2. Is the code designed for maintainability and future modifications?

NOTE

1 Bhattacharjee, S., (July 28, 2025), "Static vs. dynamic code analysis: A comprehensive guide," vFunction, https://vfunction.com/blog/static-vs-dynamic-code-analysis/#:~:text=You%20can%20perform%20static%20code,the%20code%20without%20executing%20it, and Jelodar, H., Meymani, M., Razavi-Far, R., (March 2025), "Large Language Models (LLMs) for Source Code Analysis: applications, models and datasets," Canadian Institute for Cybersecurity, Faculty of Computer Science, University of New Brunswick, Fredericton, NB, Canada, https://arxiv.org/html/2503.17502v1#:~:text=In%20a%20broader%20context%2C%20large,l anguage%20descriptions%20of%20code%20snippets, and (n.a.), (n.d.), "AI-driven code analysis: The new frontier in code security," GitLab, https://about.gitlab.com/topics/agentic-ai/ai-code-analysis/#:~:text=Pattern%20recognition%20is%20a%20core,they%20escalate%20into%20serious%20problems.

2 Figure 18.1, TensorFlow Python Program Code Example, created from an independent query by author, software available from tensorflow.org, www.tensorflow.org.

Chapter 19

Change Management and AI Systems

CHANGE MANAGEMENT DEFINED

Change management can mean many things to many people. In the field of Human Resources or Organizational Behavior, for example, change management is defined as the systematic approach to facilitating and managing the human side of organizational change by minimizing resistance, fostering engagement, and ensuring that transitions achieve their intended outcomes with minimal disruption to operations.

This text examines change management from the perspective of IT operations, systems, and applications.

In IT, change management is defined as the process of planning, coordinating, implementing, and monitoring changes affecting any production platform within IT's control.

Change management applies to all architectures, processes, tools, metrics, and documentation, as well as changes to IT services and other configuration items. Therefore, the primary purpose of change management is to ensure that potential changes to IT service components are evaluated and authorized before the change is undertaken.

CHANGE MANAGEMENT OBJECTIVES

The objectives of the change management process are to:

- Ensure that changes are made with minimum disruption to the services IT has committed to its users.
- Support the efficient and prompt handling of all changes.
- Provide accurate and timely information about all changes.
- Ensure all changes are consistent with business and technical plans and strategies.
- Ensure that a consistent approach is used for all groups who request changes.

DOI: 10.1201/9781003646501-24

- Provide additional functionality and performance enhancements to systems while maintaining an acceptable level of user services.
- Reduce the ratio of changes that need to be backed out of the system due to inadequate preparation.
- Minimize the total number of unplanned changes.
- Ensure that the required level of technical and management accountability is maintained for every change.
- Monitor the number, reason, type, and associated risk of the changes.
- Provide documentation that allows executive management to understand, at any point in time, the overall impact of the planned change.
- Minimize the bureaucratic impact of the change on the development community while still maintaining control of the environment.[1]

ELEMENTS OF CHANGE MANAGEMENT

To protect information systems and services, a formal change management system must be established to enforce strict controls over changes to all information processing facilities, systems, software, or procedures.

End-user management must formally authorize all changes before implementation and ensure that accurate documentation is maintained.

The change management process may differ between changes to IT infrastructure (which includes all hardware, system software, and network assets) and application software (which includes commercial off-the-shelf data applications and in-house developed data application software). However, the underlying requirements for managing change are the same. All IT infrastructure and application development changes are required to follow a change management process to ensure the change is approved for release and does not unknowingly add security risks to the organization's operating environment.[2]

The change management process involves multiple elements, many of which are beyond the scope of this *Handbook*.

Two specific elements are reviewed here as they are more closely related to the processes of auditing artificial intelligence (AI). Further details regarding the definition, role, and application of additional elements within change management (e.g., risk, security, and vulnerability assessment of the change, change priorities, change implementation scheduling, scheduled maintenance, software release, and fallback procedures, test/development environments, and emergency/urgent/expedited changes) are not addressed and are left to the reader's independent review and examination.

Element #1 Change Type

Various areas within IT operations may be affected by change management. There may be changes to (list not exhaustive):

- Applications
- Databases
- Hardware
- IT Configuration
- IT Infrastructure Environment
- Operating Systems
- Software
- Telecommunications.

Element #2 Risk Level

The risk to IT operations and the enterprise, as a result of changes across any of the IT elements identified in #1 may be categorized as:

- Low
- Medium
- High.

It is beyond the scope of this book to apply modifiers, definitions, financial ranges, or impact levels to these risk categories. Doing so is the responsibility of the data owner, end-user, and senior management, working with the organization's risk management professionals.

THE ROLE OF AI IN CHANGE MANAGEMENT

Although our main focus is on auditing AI, AI can play a significant role in IT change management by automating, optimizing, and enhancing the processes related to managing changes in IT systems and applications.

The following are some key roles AI may play in the change management process:

1. Automating Routine Tasks
   AI can automate repetitive and time-consuming tasks, such as change request logging, approvals, and notifications, reducing human effort and errors.
2. Predictive Analytics
   AI can predict the potential impact of proposed changes by analyzing historical data, identifying patterns, and forecasting risks or issues that may arise.
3. Risk Assessment
   AI tools can assess the risk level of changes by analyzing the complexity, scope, and dependencies of the proposed modifications, helping IT teams make informed decisions.

4. Impact Analysis
   AI can map dependencies across IT systems and applications to determine how a change in one area might affect other systems or services.
5. Incident Prediction and Prevention
   By analyzing past incidents and their triggers, AI can predict which changes are more likely to cause disruptions and suggest preventive measures.
6. Real-Time Monitoring and Feedback
   AI can monitor systems in real-time during and after a change implementation to detect anomalies and provide immediate feedback.
7. Improved Communication
   AI-powered chatbots and virtual assistants can facilitate communication between IT teams and stakeholders, ensuring that all parties are informed and involved in the change management process.
8. Knowledge Management
   AI can create and maintain a centralized knowledge base of past changes, outcomes, and best practices, enabling teams to learn from historical data.
9. Compliance and Auditing
   AI can ensure that changes comply with organizational policies and regulatory requirements by cross-referencing changes with compliance guidelines.
10. Continuous Improvement
    AI-driven insights can help organizations refine their change management processes over time, making them more efficient and resilient.

AUDITING CHANGE MANAGEMENT IN AI-BASED SYSTEMS

AI systems are not fundamentally different from non-AI systems in terms of the need for internal controls and risk monitoring. Both require rigorous controls to ensure reliability, security, and compliance with policies and regulations. However, the nature of risks and controls for AI systems can differ due to the unique characteristics of AI, such as adaptability, data dependency, and potential biases.

While the core principles of IT change management apply to both AI and non-AI systems, the specific risks and controls associated with AI require tailored approaches. By addressing the unique challenges of AI, organizations can integrate robust change management practices to ensure the reliability, security, and compliance of their AI systems.

When tasked with assessing change management policies, practices, risks, and the impact on AI systems' internal controls, the auditor or risk assessor should consider incorporating the following questions into their respective audit/assessment examination process:

Policy and Governance

1. How does the change management policy address AI-specific challenges, such as managing changes to machine-learning models or reward functions?
   Rationale: Ensures the policy is tailored to the unique nature of AI systems, which require frequent updates and testing.
2. What governance frameworks or boards (e.g., Change Advisory Boards) are in place to oversee and approve changes to AI systems, and how do they incorporate technical and ethical considerations?
   Rationale: Evaluates oversight mechanisms and inclusion of ethical and technical dimensions in decision-making.
3. What formal processes are in place to manage and approve changes to the AI system?
   Rationale: Ensures that the organization has a structured and consistent approach to managing changes.
4. Please describe the approval workflow for different types of changes.
   Rationale: Verifies that clear and appropriate approval mechanisms exist based on the type and impact of changes.
5. How does the organization ensure that changes to the AI system comply with regulatory requirements and internal policies?
   Rationale: Evaluates alignment with external regulations and internal governance standards.
6. What steps are taken to align changes with legal and regulatory frameworks?
   Rationale: Ensures compliance with evolving legal and regulatory requirements.

Risk Assessment and Impact Analysis

1. How are risks, such as unintended consequences of model updates or deployment of poorly tested algorithms, assessed and documented in the change management process?
   Rationale: Identifies processes to mitigate risks specific to AI systems.
2. What processes are in place to assess the risks and potential downstream effects of changes made to AI models, algorithms, or training datasets?
   Rationale: Ensures risks unique to AI systems are systematically identified and addressed.
3. How are changes to AI systems evaluated for their impact on key stakeholders, including regulatory bodies, customers, and internal teams?

Rationale: Examines the organization's ability to analyze the broader effects of AI system changes.

4. How are changes to the AI system, including updates, patches, and modifications, identified and classified?

   Rationale: Ensures changes are systematically identified and categorized to evaluate potential risks.
5. What criteria are used to determine the significance of a change?

   Rationale: Establishes the organization's framework for assessing the impact and priority of changes.
6. How does the organization ensure that changes to the AI system do not negatively impact system performance or introduce new risks?

   Rationale: Evaluates risk mitigation processes to ensure performance stability and security.

Testing and Validation

1. What testing and validation procedures are performed before deploying changes to AI systems, and how do they ensure alignment with original objectives?

   Rationale: Validates robustness and alignment of system changes.
2. How does the change management process ensure separate testing, staging, and production environments for AI systems?

   Rationale: Promotes safe deployment practices and avoids unintended production impacts.
3. What testing protocols are implemented to validate AI system changes, and how do these protocols ensure that the system continues to meet performance, fairness, and security benchmarks?

   Rationale: Evaluates the robustness of testing processes for AI systems.
4. How is the integrity of AI system outputs verified following changes, particularly when training data or reward functions are updated?

   Rationale: Ensures that changes do not compromise system integrity or introduce unintended outcomes.
5. What testing and validation steps are taken before implementing changes?

   Rationale: Ensures the organization verifies the safety, reliability, and effectiveness of changes before deployment.
6. How are changes to AI models specifically handled, including retraining, tuning, or replacing models?

   Rationale: Assesses specific processes for managing updates or modifications to AI models.
7. What criteria determine whether a model needs to be retrained or replaced?

   Rationale: Verifies the existence of clear guidelines to evaluate when model updates are required.

Documentation and Transparency

1. What documentation practices are followed to track the lifecycle of change requests for AI systems, including approvals, testing outcomes, and deployment logs?
   Rationale: Ensures traceability and accountability for changes.
2. How does the organization handle transparency in documenting changes to sensitive components, such as training datasets or algorithm updates?
   Rationale: Addresses concerns related to bias, data integrity, and transparency.
3. What mechanisms are in place to document all changes to AI systems, including modifications to training datasets, algorithms, and reward structures, and how is this information shared with relevant stakeholders?
   Rationale: Promotes transparency and accountability in change management.
4. How are change requests and their justifications for AI systems tracked and stored, and what role does this documentation play in compliance audits?
   Rationale: Ensures change records are accessible for future reviews and compliance purposes.
5. What documentation is maintained for each change made to the AI system, and how is it updated and reviewed?
   Rationale: Ensures that changes are properly documented for accountability, traceability, and auditability.
6. How do you ensure the documentation remains accurate and up-to-date?
   Rationale: Evaluates processes for maintaining documentation accuracy in a dynamic environment.
7. How is communication regarding proposed/completed changes handled across different departments?
   Rationale: Assesses coordination and communication mechanisms to ensure all stakeholders are informed.

Contingency Planning and Rollbacks

1. What contingency plans and rollback procedures are established for AI systems, especially when model or algorithm updates lead to unexpected behaviors or failures?
   Rationale: Examines preparedness for mitigating risks associated with AI-specific changes.
2. How are rollback procedures defined and executed in the event of an unsuccessful change to the AI system?
   Rationale: Ensures there are robust fallback plans to minimize disruption in case of failed changes.

3. Please provide examples of when rollback procedures have been used.
   Rationale: Assesses the effectiveness and practical application of rollback strategies.

Ongoing Monitoring and Continuous Improvement

1. How is the performance of AI systems monitored post-change, and what feedback mechanisms exist to refine future change management processes?
   Rationale: Ensures continuous improvement and responsiveness to system performance.
2. What mechanisms are in place to monitor the impact of changes on the AI system after implementation?
   Rationale: Ensures ongoing evaluation to detect unintended consequences and optimize performance.
3. What tools and metrics are used to continuously monitor the effectiveness and impact of changes to AI systems over time?
   Rationale: Ensures that the organization employs reliable tools and metrics to assess the long-term success of implemented changes.
4. How are insights from post-implementation monitoring used to inform and refine future change management practices for AI systems?
   Rationale: Evaluates how the organization leverages monitoring data to improve processes and reduce risks in future changes.

Training and Support

1. What training and support are provided to users and administrators to adapt to changes in the AI system?
   Rationale: Ensures stakeholders can effectively adapt to system changes without productivity losses.
2. How is change-related training documented and evaluated?
   Rationale: Assesses processes for ensuring the quality and effectiveness of training initiatives.
3. How are training materials tailored to address the unique needs of different user groups, such as administrators, developers, and end-users?
   Rationale: Ensures that training content is customized to meet the specific roles and responsibilities of various stakeholders, maximizing its effectiveness.
4. What feedback mechanisms are in place to evaluate and improve the training provided for changes to AI systems?
   Rationale: Validates the organization's commitment to continuous improvement in training programs based on user input and outcomes.

Stakeholder Involvement

1. How are stakeholders, including business units and IT, involved in the change management process for AI systems?

Rationale: Evaluates the organization's inclusivity and stakeholder engagement in decision-making and implementation.

2. How are stakeholders' concerns and requirements captured and prioritized during the change management process for AI systems?

   Rationale: Ensures that stakeholder input is systematically considered and effectively integrated into the change process.

3. What strategies are used to maintain stakeholder engagement throughout the lifecycle of a change, from proposal to post-implementation?

   Rationale: Evaluates the organization's approach to sustaining collaboration and communication with stakeholders to ensure successful change adoption.

NOTES

1 Information Technology Department, February 15, 2022, "Change Management Process," University of Cincinnati, www.uc.edu/content/dam/uc/ucit/docs/so/cm_process_document.pdf.

2 Nebraska Information Technology Commission, July 14, 2023, "Technical Standards and Guidelines, 8-202. Change control management," https://nitc.nebraska.gov/standards/index.html.

Chapter 20

Safety and Security of AI Systems and Software

The Blueprint for an AI Bill of Rights: Making Automated Systems Work for the American People was published by the White House Office of Science and Technology Policy (OSTP) in October 2022.

The OSTP's Blueprint for an AI Bill of Rights identified five principles that should guide the design, use, and deployment of automated systems to protect the American public in the age of artificial intelligence (AI). As its first value, this AI Bill of Rights calls for establishing safe and effective systems by setting forward the following individual "right:"

> You should be protected from unsafe or ineffective systems. Automated systems should be developed with consultation from diverse communities, stakeholders, and domain experts to identify concerns, risks, and potential impacts of the system. Systems should undergo pre-deployment testing, risk identification, and mitigation, and ongoing monitoring that demonstrate they are safe and effective based on their intended use, mitigation of unsafe outcomes including those beyond the intended use, and adherence to domain-specific standards.[1]

AUDITING THE SAFETY OF AI SYSTEMS AND SOFTWARE – GENERAL QUESTIONS

The following questions should be asked regarding the safe and effective use of an AI tool by the organization:

1. Are there measures in place to minimize negative consequences caused/created by the AI tool that would directly affect/impact the organization?
2. Can the AI tool be controlled or constrained to operate within defined safety limits? How are these controls substantiated?
3. Is there redundancy or fail-safe mechanisms in place in case the AI tool malfunctions? Have these been substantiated? Tested?

DOI: 10.1201/9781003646501-25

4. What are the potential failure modes of the AI tool, and how are they mitigated?
5. What is the potential impact of a failure or erroneous decision by the AI tool on decisions rendered or relied upon by the organization's personnel, clients, and the organization in general?
6. What mechanisms exist to prevent/mitigate the AI tool from making extreme or harmful decisions? How are these mechanisms tested? Documented?
7. What testing procedures have been conducted to validate the AI tool's performance and safety? How have these been substantiated?

Executive management should be acutely aware that as the organization adopts and deploys AI technology (in various forms and across multiple firm computer systems), threat actors are also employing AI.

Threat actors are using AI and adapting their methods to incorporate AI tools. Threat actors are using AI tools to evade detection and gain access to the organization's IT systems, applications, and data. The increased use of AI tools as a means to thwart established, internal cybersecurity controls exponentially increases the risk exposure to the organization's digital assets, and the organization's clients, and impacts the organization's overall business resiliency capabilities.

Establishing and sustaining internal controls designed to mitigate the risks associated with AI systems/tools should build upon the organization's existing IT internal controls. Management should treat the implementation of AI tools as another albeit advanced technology that will require continual monitoring, adaptation of existing controls, implementation of additional, new controls, and adherence to internal control policies and procedures.

Management should consider bolstering IT security controls in these key areas with specific relevance to all AI tools acquired and deployed (or to be acquired) by the organization.

The review of these key areas may be conducted by the organization's internal IT audit function (if present). If the organization does not have an internal IT audit function or if the skill set of personnel is not sufficient to examine AI systems, because portions of this security review require specific skill sets, then contracting an external, qualified third party to perform these reviews is both logical and a cost-effective alternative.

The following should be addressed when auditing the security of AI software:

1. Review data privacy and protection controls and how the AI software achieves both.
2. Examine the AI software's robustness and approach to bias mitigation.
3. Assess the security of components of the software, and control that mitigate unauthorized modifications of these components.

4. Conduct both adversarial testing and penetration testing of the AI software.
5. Assess the software's capability of meeting the requirements of explainability and accountability.[2]

AUDITING THE SECURITY OF AI SYSTEMS AND SOFTWARE – GENERAL QUESTIONS

The following questions establish a basis for developing a more comprehensive general IT audit controls questionnaire for the audit of the security of AI systems and software, acquired and deployed by the organization.

1. Can humans intervene to take control if the AI tool's behavior becomes unpredictable or risky?
2. Can you describe the security measures implemented to safeguard the AI system from cyber threats, such as hacking or malware?
3. Can you provide examples of how security incidents related to the AI system have been handled and what lessons were learned?
4. How does the organization ensure secure integration of AI systems with existing IT infrastructure and other applications?
5. How does the organization ensure that AI software complies with data privacy regulations (e.g., GDPR, CCPA) when processing sensitive information?
6. How does the organization ensure that its AI systems comply with relevant laws, regulations, and industry standards?
7. How does the organization ensure that the data used for training AI models is accurate, complete, and free of biases?
8. How does the organization evaluate and manage risks associated with third-party AI tools, libraries, or cloud services?
9. How does the organization evaluate the performance of its AI systems post-deployment, and what mechanisms are in place to incorporate improvements?
10. How does the organization manage changes to AI software, including updates and retraining of models?
11. How does the organization safeguard AI models against adversarial attacks or unauthorized modifications?
12. How does the organization detect and respond to potential vulnerabilities in the AI system?
13. How does the organization ensure that the AI system is protected against unauthorized access, and what specific controls are in place?
14. How does the organization ensure that third-party vendors or partners involved with the AI system, used by the organization, adhere to the same security standards employed in other enterprise-wide systems?

15. How does the organization manage access controls and authentication mechanisms for individuals interacting with the AI system?
16. How is feedback from end-users of AI systems collected and utilized to identify potential security vulnerabilities or functional issues?
17. How is the performance and security of deployed AI software monitored, and what logging mechanisms are in place?
18. Is there a mechanism to quickly shut down the AI tool in case of emergencies or unexpected behavior?
19. What documentation and reporting practices are in place to ensure transparency and accountability in the security management of the AI system?
20. What is the organization's plan for responding to security incidents involving AI systems?
21. What measures are in place to ensure that AI systems continue to operate reliably under various conditions, including potential failures?
22. What measures are in place to ensure the confidentiality, integrity, and availability of the data processed by the AI system?
23. What measures are in place to restrict access to AI software, its models, and associated data to authorized personnel only?
24. What policies and controls are in place to address the ethical implications of deploying AI systems?
25. What processes are in place to detect and address unintended biases in AI models during development and after deployment?
26. What processes are in place to regularly assess and update the security protocols for the AI system?
27. What risk assessment process is followed before deploying AI systems, and how are identified risks mitigated?
28. What steps are taken to document and maintain transparency about how AI algorithms make decisions?
29. What steps are taken to secure the AI system's data pipelines and ensure that data is transmitted and stored securely?
30. What training and awareness programs are in place for employees interacting with AI systems?

NOTES

1 White House Office of Science and Technology Policy, October 2022, "Blueprint for an AI BILL OF RIGHTS making automated systems work for the American People," The White House, https://bidenwhitehouse.archives.gov/ostp/ai-bill-of-rights/.

2 Marcella, A., and Renz, G., January 2024, "Generative Artificial Intelligence: Benefits and Risks to Law Firms," Association of Legal Administrators, https://www.alanet.org/publications/white-papers/generative-artificial-intelligence-benefits-and-risks-to-law-firm.

Chapter 21

Software Certification of AI Systems

DEFINING CERTIFICATION

Generally speaking, certification in software development is a process by which a software developer acquires attestation of compliance with a quality standard.

Certification is a confirmation by a "third party" that requirements of, e.g., international standards, industry specifications, or technical rules are met. Certification is based on a conformity assessment in which the fulfillment of the requirements is checked. The subject of such assessments can be, for example, products, projects, processes, or management systems.

The word certification comes from the Latin phrase "certus facere," literally "to make something safe." "Making something safe" is expressed in a confirmation or declaration that a conformity assessment has shown compliance with certain requirements, for example, by way of an external audit. The externally visible proof of certification is the corresponding certificate.[1]

ATTESTATION VS CERTIFICATION

Attestation refers to a formal declaration by an organization, asserting that its software development processes adhere to specific policies, frameworks, or guidelines, often without external verification.

Certification, on the other hand, involves an independent third-party assessment to validate compliance, resulting in an official certification or seal of approval. While attestation relies on self-reporting, certification provides externally validated assurance of compliance.

SOFTWARE CERTIFICATION AND CERTIFIED SOFTWARE

Software certification demonstrates the reliability and safety of software systems in such a way that it can be checked by an independent authority

DOI: 10.1201/9781003646501-26

with minimal trust in the techniques and tools used in the certification process itself. It builds on existing software assurance, validation, and verification techniques but introduces the notion of explicit software certificates, which contain all the information necessary for an independent assessment of the demonstrated properties.

Software certification comprises a wide range of formal, semi-formal, and informal assurance techniques, including formal verification of compliance with explicit safety policies, system simulation, testing, code reviews, human "sign-offs," and even references to supporting literature. Consequently, the certificates can have different types, and the certification process requires different mechanisms.[2]

Zhong Shao provides a definitive definition of certified software stating that certified software consists of a machine-executable program paired with a formal, machine-checkable proof that verifies the software is free of bugs concerning specific dependability claims. It emphasizes rigorous formal specifications and proofs to ensure dependability.[3]

Why Is Software Certification Important?

Certifying application software is important because it provides a verified guarantee of quality, security, and compliance with industry standards, which enhances customer confidence, improves market standing, and can lead to better product adoption by demonstrating that the software meets specific requirements and is reliable, especially in sectors with strict regulations; essentially, it signifies that the software has been rigorously tested and adheres to best practices, boosting its credibility and value to potential users.

The following is a list of ten key reasons why the certification of application software matters.

1. Ensures Compliance
   Certification verifies that software adheres to industry standards, regulations, and legal requirements, reducing the risk of penalties or legal challenges.
2. Builds Trust and Credibility
   A certified software application demonstrates reliability and quality, fostering trust among users, stakeholders, and customers.
3. Enhances Security
   Certification often includes rigorous security evaluations, ensuring the software is resilient to vulnerabilities and cyber threats.
4. Promotes Ethical Practices
   Certified applications are assessed for adherence to ethical guidelines, such as fairness, transparency, and accountability, particularly for artificial intelligence (AI)-based systems.

5. Improves Market Competitiveness
   Certification serves as a differentiator in the marketplace, signaling that the software meets recognized standards and instills confidence in prospective customers.
6. Minimizes Risk
   The certification process identifies and addresses potential risks in software, from functionality and usability issues to regulatory non-compliance.
7. Supports Interoperability
   Certified software is often validated for compatibility with other systems and applications, ensuring seamless integration in diverse environments.
8. Facilitates Global Acceptance
   Certifications based on international standards (e.g., ISO, IEC) ensure the software is recognized and accepted across global markets.
9. Encourages Continuous Improvement
   Certification processes require periodic re-evaluation, encouraging developers to maintain high-quality standards throughout the software lifecycle.
10. Protects End-Users
    Certification ensures that applications are safe, reliable, and functional, minimizing risks to end-users and enhancing user experience.

Certified software addresses challenges such as software complexity, lack of dependability metrics, and system software vulnerabilities. It aims to transform software from being a system's weakest link into one of its most dependable components.[4]

STANDARDS FOR SOFTWARE DEVELOPMENT VS CERTIFYING A SPECIFIC SOFTWARE APPLICATION

The following material provides a discussion of the subtleties between software development standards and the certification of specific software applications. Highlighted for each are the distinct scopes, purposes, and focuses of standards as frameworks guiding the development process and certifications as evaluations of individual software compliance.

Standards for Software Development

- Scope: These are general frameworks, best practices, and guidelines that outline how software should be designed, developed, and deployed. They apply broadly to the software development lifecycle and aim to ensure quality, consistency, and reliability.
- Purpose: Standards provide developers and organizations with a shared foundation to create software that meets industry-wide technical, ethical, and operational benchmarks.

- Focus: Standards focus on processes, methodologies, and tools for software creation. For example: ISO/IEC 42001:2023 focuses on AI management system requirements. Agile or DevOps methodologies focus on iterative, collaborative software development.
- Example: Following coding standards (e.g., secure coding practices) during the development phase.

Certifying a Specific Software Application

- Scope: Certification evaluates whether a particular software product complies with specific standards, regulatory requirements, and ethical guidelines, particularly for responsible system design, development, and deployment.
- Purpose: The goal is to validate that the individual software application meets defined criteria for responsible practices, ensuring it operates as intended, is trustworthy, and aligns with legal, ethical, and societal expectations.
- Focus: Certification focuses on the outcome – assessing the software's compliance with criteria like transparency, fairness, security, and maintainability, as well as demonstrating that these practices were followed during development.
- Example: Certifying that an AI-based hiring platform ensures bias-free decision-making by adhering to fairness metrics, auditing processes, and ethical data use as outlined by frameworks like the EU AI Act or NIST AI RMF.

In summary, standards act as a roadmap for building responsible software, while certification evaluates whether a specific product has successfully adhered to those principles and meets the expectations for responsible deployment and use (see Table 21.1).

Table 21.1 Comparison of Software Development Standards and Application Certification

Aspect	*Standards for Software Development*	*Certifying a Specific Software Application*
Objective	Guide development processes and practices.	Validate compliance with specific criteria for responsible practices.
Application	Broadly applies to all software development projects.	Focuses on assessing an individual software product.
Focus	Emphasis on methodologies and development best practices.	Emphasis on outcomes and adherence to ethical, legal, and technical criteria.
Output	Produces software aligned with industry standards.	Issues a certification or label indicating the software's compliance.
Examples	Use of secure coding guidelines or Agile development frameworks.	Certification under ISO/IEC 42001:2023 or EU AI Act compliance.

GLOBAL STANDARDIZATION BODIES – SOFTWARE CERTIFICATION

The following is a representative list of standardization bodies that develop standards and certify software to ensure software products are reliable, secure, and compliant with international and local regulations. Reviewing these organizations and their respective standards will provide additional resources for the design of internal audit and assessment questions applicable to AI systems (see Table 21.2).

- AFNOR Certification (France)
  AFNOR governs the NF Certification, which ensures software complies with French and European standards.
- ANSI (American National Standards Institute)
  ANSI accredits standards and certifies software under frameworks aligned with US and international regulations. ANSI does not develop its own standards but accredits organizations and standards developed by other bodies, often in collaboration with international entities like ISO and IEEE.
- BSI (British Standards Institution)
  BSI provides certification for software based on international and British standards, including ISO/IEC frameworks. BSI Certification for IT Application Software focuses on Certification against ISO 9001, which ensures that software development processes follow consistent and high-quality practices. ISO/IEC 27001 to ensure compliance with information security management practices, critical for software handling sensitive data, and ISO/IEC 20000 demonstrating excellence in IT service management for software solutions.
- DIN (Deutsches Institut für Normung e.V.)
  DIN develops German standards for software certification and works in coordination with ISO and IEC. Systems and software engineering – Quality criteria and evaluation of systems and software products (SQuaRE) – Common industry format for usability – Evaluation report (ISO/IEC 25066:2016); German version EN ISO/IEC 25066:2019.
- IEC (International Electrotechnical Commission)
  IEC works with ISO to develop standards for software and IT systems, particularly in electrotechnology and automation.
- IEEE (Institute of Electrical and Electronics Engineers)
  IEEE provides standards for software systems, such as IEEE 730 (Software Quality Assurance) and IEEE 1012 (Software Verification and Validation).
- ISO (International Organization for Standardization)
  ISO develops international standards for software quality, including ISO/IEC 25000 System and Software Quality Requirements and

Evaluation (SQuaRE) framework to evaluate software product quality and ISO/IEC 42001 promoting the responsible development and use of AI systems.

- TÜV SÜD (Technischer Überwachungsverein, Germany)
  TÜV SÜD certifies software for safety, reliability, and compliance in critical sectors like automotive and healthcare.

Table 21.2 Global Software Certification Agencies and Organizations

Organization	*Website*
AFNOR Certification (France)	www.afnor.org
ANSI (American National Standards Institute)	www.ansi.org
BSI (British Standards Institution)	www.bsigroup.com
CISQ (Consortium for IT Software Quality)	www.it-cisq.org
CMMI Institute	www.cmmiinstitute.com
DIN (Deutsches Institut für Normung e.V.)	www.din.de
ENISA (European Union Agency for Cybersecurity)	www.enisa.europa.eu
IEC (International Electrotechnical Commission)	www.iec.ch
IEEE (Institute of Electrical and Electronics Engineers)	www.ieee.org
IREB (International Requirements Engineering Board)	www.ireb.org
ISO (International Organization for Standardization)	www.iso.org
NIST (National Institute of Standards and Technology)	www.nist.gov
SEI (Software Engineering Institute)	www.sei.cmu.edu
TÜV SÜD (Technischer Überwachungsverein)	www.tuvsud.com

CERTIFYING AI SOFTWARE

In his paper, "Qualifying and Regulating the Use of Artificial Intelligence in Software Systems for the Canadian Nuclear Sector," Samer Dahaweer proposed three specific methodologies that would assist in building a qualification framework for AI software.

These three specific methodologies are summarized as follows:

1. Model Interpretability Analysis – Ensuring explainability of AI decision-making to allow stakeholders to be able to trust the process and hold those involved accountable.
2. Feature Importance Analysis – Which variables are the most relevant to predicting the model's output? That is, which inputs are most important to make AI behave as it does?

3. Data Diversity Analysis – Ensuring a diverse and representative range of data is used in training, reducing biases, and improving generalizability across various scenarios.[5]

These three methodologies were proposed primarily because they are central aspects of assessing AI model performance, relating directly to safety, reliability, and regulatory compliance. These, as previously discussed, are essential core criteria required for software certification.

A further examination of the three methodologies proposed by Dahaweer is presented below to demonstrate linkages between the proposed methodologies, the ability of the methodologies to assist in building a qualification framework of AI software, and the logical next step in adapting these methodologies into an acceptable framework to use in certifying AI software.

What Is AI Interpretability?

When all is said and done, AI interpretability is the ability of a carbon-based entity (we humans) to understand and explain the decision-making processes that drive AI models.

AI models use a complex web of data inputs, algorithms, logic, data science, and other processes to return insights. The more complex the model, the more difficult it can be for humans to understand the steps that led to its insights – even if those humans are the ones who designed and built it.

An interpretable model is one whose decisions can be easily understood by users.[6]

White-Box Models vs Black-Box Models

Black-box and white-box models are two model types used in machine-learning and AI systems that differ in how transparent they are about their inner workings.

Black-box models in machine-learning and AI are systems where the internal decision-making processes are opaque or difficult to interpret. These models, such as deep neural networks, focus on achieving high accuracy but lack transparency, making it hard to understand the steps in the AI system's decision-making process and how specific decisions are reached. Users know the input and output but cannot see or explain the reasoning behind the decisions.

White-box (sometimes called glass-box) models are machine-learning and AI systems that are fully transparent, allowing users to understand how decisions are made. These models, like linear regression or decision trees, provide interpretable outputs where the relationship between inputs and outputs is clear and easily explainable. White-box models prioritize transparency and interpretability.

Feature Importance

Feature importance is a technique used in machine learning that assigns a score to input features based on how useful they are in predicting a target variable. These scores provide a ranking, allowing you to understand which features are most influential in your model's prediction.

Feature importance illuminates the relationship between the features and the target variable, providing valuable insights into your data. It is a useful tool for machine learning and data analysis and is a key component in explainable AI (XAI) strategies.[7]

Feature importance analysis has many applications in diverse areas, including local climate zone analysis, patient management, short-term load forecasting, tornado prediction, and house price prediction as examples.

Example of Feature Importance: Predicting Student Academic Success

Imagine an academic institution wants to build a machine-learning model to predict whether a student will pass or fail a course. The model uses various features (inputs) such as:

- Attendance rate (e.g., percentage of classes attended),
- Study hours per week,
- Participation in class activities,
- Previous GPA, and
- Assignment submission rate.

Once the model is trained, it determines that attendance rate and study hours per week are the most important features in predicting a student's success. This means these two features have the most significant impact on whether the model predicts that a student will pass or fail.

By identifying feature importance, the academic institution can focus on improving these factors (e.g., encouraging regular attendance and study habits) to help students succeed.

This example shows how feature importance helps explain which factors are most influential in the model's decision-making process.

Why Is Feature Importance Useful in Machine Learning?

Feature importance is useful in machine learning because it helps identify which inputs (features) have the greatest impact on a model's predictions. By understanding the importance of each feature, developers and decision-makers can focus on the most relevant factors, improving the model's efficiency and accuracy.

It also enhances transparency, allowing users to interpret the model's decisions and ensure fairness. Additionally, it can guide data collection efforts, reducing unnecessary complexity by prioritizing key features.

Data Diversity

> Getting a second opinion is a common practice for humans — it must be true for automation, too. We need data diversity.
>
> *Jennifer Belissent, Principal Data Strategist, Snowflake, Inc.*

A well-designed machine-learning system typically comprises three key components: plentiful training data, an effective model training process, and accurate inference.

While numerous factors influence the performance of machine learning, one stands out as particularly vital – the diversity of the training data. Emphasizing diversity throughout the machine-learning process ensures the development of robust and high-performing models of different tasks.[8]

Data diversity in AI systems is defined as the inclusion of a broad range of data types, sources, and representative examples, covering different classes, categories, situations (scenarios), and samples from various groups to ensure the AI system performs well across different populations and scenarios. The objective is to avoid bias and "hallucinating" (i.e., making mistakes) and improve the AI system's generalizability, ensuring the system accounts for variability in real-world data.

Example:

> A facial recognition system trained on a diverse dataset containing images of people from different ethnicities, age groups, and lighting conditions will be more accurate and fairer in identifying individuals from all backgrounds.

Data diversity is important because it is one factor contributing to assuring that AI systems are fair, unbiased, and perform accurately for all users, reducing the risk of discrimination and improving trust and reliability in the system's decisions.

RESPONSIBLE AI FOUNDATION PRINCIPLES, FRAMEWORKS, STANDARDS, GUIDELINES, AND LEGISLATION

One of the most important uses of having standards to measure software quality characteristics is to be able to certify software to be at acceptable levels of risk to the business or mission the software supports.[9]

An examination of the following principles, frameworks, standards, guidelines, and legislation provides a foundation for establishing core criteria that an organization may use to assess, validate, demonstrate, and yes, eventually certify that the organization has implemented and adheres to responsible AI system design, development, and deployment practices. It should be noted that the following list is representative and not exhaustive, and is broadly representative. There are additional principles, frameworks, standards, guidelines, and legislation that lend themselves to assessing responsible AI, and those represented in the list below are most noteworthy.

Responsible AI Foundation Principles, Frameworks, Standards, Guide lines, and Legislation

- EU AI Act
- Google's AI Principles (Updated 2023)
- Hiroshima AI Process Comprehensive Policy Framework
- IBM – AI Ethics and Trust Framework
- ISO/IEC 23894:2023 Artificial Intelligence – Guidance on Risk Management
- ISO/IEC 42001:2023 Artificial Intelligence Management System Standard
- ISO/IEC 5338:2023 Artificial intelligence – AI System Lifecycle Processes
- ISO/IEC TR 24028:2020 Artificial Intelligence – Overview of Trustworthiness in Artificial Intelligence
- Microsoft – Responsible AI Standard (v2)
- NCSC Guidelines for Secure AI System Development
- NIST AI Risk Management Framework
- OECD AI Principles
- OpenAI – AI Safety and Alignment Standards
- The Advancing American AI Act
- The AI in Government Act of 2020
- The EU General Data Protection Regulation (GDPR)
- The White House Blueprint for an AI Bill of Rights 2022.

CORE CRITERIA FOR RESPONSIBLE AI SYSTEM DESIGN, DEVELOPMENT, AND DEPLOYMENT PRACTICES

The following represent the core criteria for assessing (and eventually certifying) that an AI system is operating responsibly and following stakeholder objectives.

- Accountability and Governance

Establish a governance framework assigning accountability for AI decisions and impacts, including monitoring, auditing, and mechanisms for addressing grievances or risks.

- Compliance with Regulations and Standards
  Demonstrate adherence to applicable laws, industry standards, and global frameworks such as ISO/IEC 42001:2023, NIST AI RMF, and the EU AI Act.
- Data Quality and Integrity
  Use robust methods to collect, process, and validate data to ensure accuracy, representativeness, and security, minimizing the risks of biased or flawed datasets.
- Ethical Alignment
  Align AI system goals and operations with ethical standards, societal values, and human rights, ensuring systems do not exacerbate inequalities or harm.
- Fairness and Equity
  Implement practices to identify and mitigate biases throughout the AI lifecycle to ensure equitable outcomes across diverse groups, avoiding discrimination.
- Human Oversight and Autonomy
  Maintain human involvement in decision-making processes for critical applications and ensure AI systems align with human values and autonomy.
- Maintainability
  Ensures that AI systems are designed to be adaptable, upgradable, and easily managed throughout their lifecycle.
- Performance Efficiency
  Confirms that AI systems operate optimally in terms of resource usage, speed, accuracy, and scalability.
- Risk Management and Mitigation
  Regularly assess risks associated with AI systems, including ethical, security, harmful bias, and societal risks, and implement mitigation strategies aligned with regulatory requirements.
- Robustness and Reliability
  Validate the technical performance of AI systems under diverse conditions, ensuring they function as intended and can withstand adversarial attacks or unexpected changes.
- Security and Privacy
  Safeguard AI systems against unauthorized access and ensure compliance with data privacy regulations, incorporating secure data handling and encryption measures.
- Transparency and Explainability
  Organizations should ensure AI systems are transparent, with clear documentation and processes that enable stakeholders to understand how the systems make decisions and how data is used.

APPLYING CORE CRITERIA TO CERTIFYING AI SOFTWARE

The foundational core criteria identified above for responsible AI provide a comprehensive framework that can serve as the basis for certifying AI software by establishing clear standards and criteria for evaluation.

The following summarizes how these core criteria may be used for certification:

- Development of Certification Criteria
  The foundational areas (e.g., transparency, fairness, security) can be translated into measurable certification criteria, detailing specific requirements that AI software must meet to be certified. For example, a system could require documented transparency mechanisms, bias detection protocols, and robust data integrity checks.
- Creation of a Certification Process
  Certification bodies can develop a multi-step process to assess AI systems against these criteria. This could include initial audits, technical evaluations, risk assessments, and interviews with development teams to validate compliance.
- Independent Auditing and Validation
  Third-party auditors could assess AI software against these standards. They would evaluate compliance in areas, such as governance, ethical alignment, and robustness, through documentation review, testing, and inspection of AI workflows.
- Ongoing Monitoring and Re-certification
  Certification could require periodic reviews to ensure continued compliance with evolving standards and regulations. Organizations would need to demonstrate updates to mitigate newly identified risks or adapt to changes in societal or regulatory expectations.
- Global Alignment and Benchmarking
  Certification schemes could leverage these foundational principles to align with international frameworks such as ISO/IEC 42001:2023 or the EU AI Act, ensuring that certified AI software is interoperable and adheres to globally recognized standards.
- Endorsement of Best Practices
  Certified AI software could be recognized as meeting best practices for responsible AI, giving stakeholders confidence in its fairness, transparency, and accountability, which could also serve as a market differentiator.
- Integration with Sector-Specific Standards
  The core criteria could be tailored to sector-specific needs (e.g., healthcare, finance, transportation) to ensure that certification accounts for unique regulatory and ethical considerations relevant to those industry sectors.

- Certification Labeling and Consumer Trust
  Certified AI software could carry a trust label or badge, signaling to users and stakeholders that the system meets established responsible AI standards, thereby promoting confidence and trust.

Using these foundational areas as a basis, certification programs can provide a structured, reliable, and globally recognized method to validate the ethical and responsible development of AI software. This would ensure consistency and accountability across diverse applications and industries.

Certified software ensures high reliability, improves maintainability, and enables modular verification of software components. It also supports continuous improvement in software dependability through machine-checkable proofs and rigorous formal specifications.[10]

AI SYSTEM DEVELOPMENT PROCESSES: ATTESTATION AUDIT

Internal auditors may lack full independence to perform an attestation audit because they are part of the organization they are auditing. Attestation audits often require impartiality to ensure credibility and eliminate potential conflicts of interest.

It may be acceptable for internal auditors to perform attestation audits if the results are used for internal purposes only and do not require formal third-party certification.

The following questions serve as the foundation for conducting an internal audit of AI system development processes. This internal audit represents the organization's initial step in preparing for a formal, independent, third-party assessment, with the ultimate goal of obtaining official certification for its software through rigorous examination and testing by an external, independent authority.

General Certification Requirements

Documentation and Standards Compliance

1. How does the organization ensure the AI system's design, development, and deployment adhere to relevant international standards or industry specifications?

Conformity Assessment

2. What processes are in place to facilitate an independent third-party conformity assessment of the AI system, and how is evidence of compliance documented?

Governance Framework

3. How does the organization's governance framework support certification efforts, including oversight, accountability, and decision-making structures?

Reliability and Safety

Reliability Testing

4. What testing methods have been employed to validate the reliability of the AI system, and how are results independently verified?

Safety Features

5. What mechanisms are in place to ensure the safety of the AI system under diverse operational scenarios, and how is compliance demonstrated?

Failure Recovery

6. How does the AI system handle failures or errors, and are these recovery mechanisms evaluated against technical rules or standards? If so, which specific rules or standards?

Transparency and Traceability

Traceability of Development

7. How is the AI system's development lifecycle documented to provide traceability and transparency? Is the lifecycle documented differently for third-party assessor requirements? If so, how?

Decision-Making Logic

8. How is the interpretability of the AI system's decision-making logic demonstrated and validated?

Audit Trail

9. How is a detailed audit trail maintained for all changes and updates to the AI system, ensuring transparency for certification purposes?

Fairness and Data Diversity

Bias Mitigation

10. What specific measures have been implemented to address potential biases in the training data, and how are these validated?

Data Diversity

11. How does the organization ensure that the AI system's training data is diverse and representative of its intended user base?

Fairness Metrics

12. What fairness metrics are applied to the AI system? How is adherence to these metrics verified internally? How will they be verified by third-party assessors?

Security and Privacy

Data Security

13. How does the AI system comply with data security standards, and how are these measures independently validated?

Privacy Protections

14. What privacy-preserving techniques are implemented in the AI system, and how is compliance with data protection regulations demonstrated?

Vulnerability Assessments

15. How are vulnerability assessments conducted, and how is the AI system's resistance to threats tested? How is it validated?

Performance and Maintainability

Performan00ce Metrics

16. What performance benchmarks are used to evaluate the AI system, and how is compliance with these benchmarks documented?

Maintainability Plan

17. What maintainability practices are in place to ensure the AI system remains compliant with standards over its lifecycle?

Ethical and Social Responsibility

Ethical Considerations

18. How does the organization ensure that the AI system aligns with ethical standards and societal values.
19. How is this alignment substantiated?

Stakeholder Engagement

20. How are diverse stakeholders engaged in the certification process to ensure the AI system meets societal and regulatory expectations?

Model Interpretability

Transparency of Decision-Making

21. How is the interpretability of the AI model ensured, and what methods are used to provide clear and understandable explanations of its decision-making process to stakeholders?

Validation of Interpretability Mechanisms

22. What steps are taken to validate the effectiveness of interpretability tools or techniques in accurately reflecting the AI system's logic, and how is this process documented for third-party assessment?

Feature Importance

Identification of Key Features

23. What methods are used to determine the most critical features influencing the AI system's predictions, and how are these features verified for relevance and accuracy?

Feature Contribution Transparency

24. How is the contribution of each feature to the AI model's predictions documented and communicated to ensure transparency for users and auditors?

Data Diversity

Assessment of Training Data Representation

25. How is the diversity of training data assessed to ensure it represents the intended population, and what steps are taken to address any identified gaps or imbalances?

Impact of Data Diversity on Performance

26. How does the organization measure the impact of data diversity on the AI system's performance, and how are findings used to improve fairness and accuracy?

Certification Renewal

Ongoing Certification Compliance

27. What processes are in place for renewing and maintaining the AI system's certification (once obtained)?
28. How is ongoing conformity assessed?

NOTES

1 (n.a.), (n.d.), "What is certification?" DQS Holding GmbH, www.dqsglobal.com/intl/learn/dqs-knowledge-center/what-is-certification
2 Denney E., and Fischer, B., (n.d.), "Software Certification and Software Certificate Management Systems," USRA/RIACS, NASA Ames Research Center, Moffett Field, CA 94035, USA, www.researchgate.net/publication/247570274_Software_Certification_and_Software_Certificate_Management_Systems#:~:text=Software%20certi%EF%AC%81cation%20demonstrates%20the%20reliability,in%20the%20certi%EF%AC%81cation%20process%20itself.
3 Shao, Z, December 1, 2010, "Certified Software," Communications of the ACM, https://dl.acm.org/doi/pdf/10.1145/1859204.1859226.
4 Shao, Z, December 1, 2010, "Certified Software," Communications of the ACM, https://dl.acm.org/doi/pdf/10.1145/1859204.1859226.
5 Dahaweer, S., October 2024, "Qualifying and Regulating the Use of Artificial Intelligence in Software Systems for the Canadian Nuclear Sector," Master Thesis, Dalhousie University, Halifax, Nova Scotia, https://dalspaceb.library.dal.ca/server/api/core/bitstreams/f3db5c05-3fc1-4fb8-9386-2744e1fdd121/content.
6 McGrath, A., and Jonker, A., October 8, 2024, "What is AI Interpretability?" IBM, www.ibm.com/think/topics/interpretability.
7 Kolena Editorial Team, 2024, "Feature Importance: Methods, Tools, and Best Practices,' Kolena, www.kolena.com/guides/feature-importance-methods-tools-and-best-practices-2.
8 (n.a.), August 8, 2023, "Why is Training Data Diversity Important for Machine Learning, AI" Blog - FutureBeeAI, www.futurebeeai.com/blog/why-is-training-data-diversity-important-for-machine-learning-ai.
9 CISQ™, (n.d.), "Software Certification," The Consortium for Information & Software Quality™ www.it-cisq.org/software-certification.
10 Shao, Z, December 1, 2010, "Certified Software," Communications of the ACM, https://dl.acm.org/doi/pdf/10.1145/1859204.1859226.

Chapter 22

Auditing AI System Performance

PERFORMANCE ASSESSMENT – SCOPE

A performance assessment evaluates whether the artificial intelligence (AI) system meets its intended objectives and operates effectively, reliably, and efficiently within its designed context.

This includes assessing the accuracy, robustness, scalability, and ethical implications of the AI model and its outputs, as well as its compliance with regulatory, operational, and organizational standards. The process typically involves reviewing model outcomes, metrics, and alignment with business goals while identifying any gaps or risks in performance.

PERFORMANCE ASSESSMENT – OBJECTIVE

The objectives of an AI performance assessment as part of an internal controls audit of an organization's AI systems should include:

- **Accuracy and Reliability:** Verify that the AI system consistently produces correct and reliable results based on predefined performance metrics (e.g., precision, recall, F1-score).
- **Bias and Fairness:** Assess the system for potential biases in its predictions and ensure fairness across different demographic or stakeholder groups.
- **Robustness and Stability:** Confirm the AI model's resilience to adversarial conditions, unexpected inputs, or environmental changes.
- **Scalability and Efficiency:** Evaluate the system's ability to handle large-scale data or increased user demands without significant degradation in performance.
- **Alignment with Goals:** Ensure the AI's outputs align with organizational objectives and ethical principles.
- **Explainability and Transparency:** Determine whether the system's decision-making process can be explained in a way that stakeholders and regulators understand.

DOI: 10.1201/9781003646501-27

- **Regulatory and Policy Compliance:** Verify adherence to legal, ethical, and organizational policies governing AI use.
- **Operational Monitoring:** Assess mechanisms for ongoing monitoring and logging of the AI system to identify issues, track performance, and implement improvements.

GAO ARTIFICIAL INTELLIGENCE ACCOUNTABILITY FRAMEWORK – AUDITING AI SYSTEM PERFORMANCE

The US Government Accountability Office (GAO), in developing the Artificial Intelligence Accountability Framework, created a set of specific audit questions aimed at assessing AI Performance.

The objectives of these questions are to:

- Catalog model and non-model components, along with operating specifications and parameters.
- Define performance metrics that are precise, consistent, and reproducible.
- Assess the performance of each component against defined metrics to ensure it functions as intended and is consistent with program goals and objectives.
- Assess whether the outputs of each component are appropriate for the operational context of the AI system.
- Document the methods for assessment, performance metrics, and outcomes of the AI system to provide transparency over its performance.
- Define performance metrics that are precise, consistent, and reproducible.
- Assess performance against defined metrics to ensure the AI system functions as intended and is sufficiently robust.
- Identify potential biases, inequities, and other societal concerns from the AI system.
- Define and develop procedures for human supervision of the AI system to ensure accountability.

Auditing AI Performance – GAO AI Accountability Framework Questions

The following is a list of audit questions designed to assess an organization's ability to meet the above AI Performance objectives.

1. How is each model component solving a defined problem?
2. How are model and non-model components' operating specifications and parameters selected, evaluated, and optimized?

3. How suitable are the components to the available data and operating conditions?
4. To what extent are the dimension reduction techniques applied appropriately?
5. What metrics has the organization developed to measure the performance of various components?
6. What is the justification for the metrics selected?
7. Who is responsible for developing the performance metrics?
8. To what extent do the metrics provide an accurate and useful measure of performance?
9. To what extent are the metrics consistent with goals, objectives, and constraints?
10. How is the selection of the mathematics and/or data science techniques, including any ensemble method, documented and assessed?
11. To what extent is the selection of techniques appropriate?
12. How appropriate is the training and optimization process for each component within the system?
13. How did the organization determine whether the output of each component is suitable for the operational context?
14. To what extent is the output of each component appropriate for the operational context?
15. To what extent are the model outputs consistent with the organization's values and principles to foster public trust and equity?
16. To what extent has the organization documented the AI system's development, testing methodology, metrics, and performance outcomes?
17. To what extent does the documentation describe test results, limitations, and corrective actions, including efforts to minimize undesired effects in the outcomes?
18. What metrics has the organization developed to measure performance of the AI system?
19. What is the justification for the metrics selected?
20. Who is responsible for developing the performance metrics?
21. To what extent are the metrics consistent with system goals, objectives, and constraints, including ethical and compliance considerations?
22. What testing, if any, has the organization conducted on the AI system to identify errors and limitations (i.e. adversarial or stress testing)?
23. Who is responsible for testing the AI system?
24. To what extent can users or parties affected by the outputs of the AI system test the AI system and provide feedback?
25. To what extent does the AI system perform differently when using different demographics or populations?
26. Which population(s) does the AI system impact?
27. Which population(s) does the AI system not impact?
28. How did the organization address disparate impacts resulting from the AI system, if any?

29. To what extent are the established procedures effective in mitigating bias, inequity, and other concerns resulting from the system?
30. How has the organization considered an appropriate degree of human involvement in the automated decision-making processes?
31. What procedures have been established for human supervision of the AI system?
32. To what extent has the organization followed its procedures for human supervision to ensure accountability?[1]

AUDITING AI ADVERSARIAL ROBUSTNESS

What Is AI Adversarial Robustness?

First, it is important to define what is meant by robustness, from both technological and AI perspectives.

The Canadian Centre for Cyber Security approaches robustness as a characterization of the security strength and assurance of a security control. The security strength is related to the control's potential ability to protect the confidentiality, integrity, or availability of IT assets. The security assurance of a control is related to confidence that the control is designed and implemented correctly and is operating as intended. Together, these two aspects define the requirements that must be met in the implementation of a control to satisfy its security objective.

For example, a conceptually strong security control (e.g., an encryption algorithm using the Advanced Encryption Standard) but comes with no assurance (e.g., there is no evidence to show that the algorithm is coded correctly) will have a lower effective robustness than a similar control that does have higher assurance (e.g., the software has been validated).

Security controls that protect more sensitive or critical IT assets or that are exposed to more significant threats will generally require stronger security solutions and require more assurance in their implementation and therefore require higher levels of robustness.[2]

Lecuyer et al. state that intuitively, a predictive model may be regarded as robust to adversarial examples if its output is insensitive to small changes to any plausible input that may be encountered in deployment. To formalize this notion, Lecuyer et al. contend that first one must establish what qualifies as a plausible input. This is difficult as academic literature, research, and industry have yet to standardize on such a definition. Instead, model robustness is typically assessed on inputs from a test set that are not used in model training – similar to how accuracy is assessed on a test set and not a property on all plausible inputs.[3]

Using the above definitions and perspectives as a conceptual and logical basis, adversarial robustness in AI refers to the ability of a machine-learning model to maintain its performance and integrity when faced with adversarial inputs – intentionally crafted examples designed to deceive or mislead the model. Robustness is characterized by the model's insensitivity to small, carefully designed perturbations in plausible inputs that aim to exploit its vulnerabilities.

Formally, adversarial robustness can be defined as the capacity of a predictive model to produce consistent and accurate outputs when subjected to adversarial perturbations within a defined space of plausible inputs. This encompasses both the strength of the model (its theoretical ability to resist adversarial attacks) and the assurance that it has been designed, trained, and evaluated in a manner that demonstrates and verifies this resistance.

Assessing adversarial robustness typically involves testing the model on adversarially generated examples, which may not fully encompass all possible perturbations but provide empirical evidence of the model's ability to withstand such attacks under deployment conditions.

Kotyan and Vargas aptly identify that, as it turns out, a simple robustness assessment is a daunting task, given the vast number of possibilities and definitions along with their exceptions and trade-offs. Improvements in robustness should also result in learning systems that can better reason over data and achieve a new level of abstraction. A robustness assessment would thus help check failures for both reasoning and high-level abstractions.[4]

Auditing AI Adversarial Robustness – Internal Control Questions

The following questions provide a starting point for the auditor or risk assessor tasked with assessing an AI system's adversarial robustness. This list is representative of questions designed to assess internal controls and risks and is not presented as being comprehensive or complete.

Model Design and Vulnerabilities

1. How was the model's architecture designed to mitigate vulnerabilities to adversarial attacks, such as evasion or model extraction?

Threat Modeling

2. What adversarial threats have been identified for the AI system, and how are they prioritized based on likelihood and potential impact?

Training Data Integrity

3. What measures are in place to ensure the integrity and security of the training data, and how are potential adversarial poisoning attacks mitigated?

Defense Mechanisms

4. What specific defense mechanisms, such as adversarial training or input sanitization, have been implemented to enhance robustness?

Evaluation Metrics

5. How is adversarial robustness evaluated, and what metrics or benchmarks are used to measure the model's resistance to attacks?

Test Data and Scenarios

6. What types of adversarial scenarios or examples have been used to test the model's robustness, and what were the results?

Model Sensitivity

7. How has the model's sensitivity to input perturbations been analyzed, and what are the thresholds for acceptable performance degradation?

Input Validation

8. What processes are in place to validate or preprocess inputs to detect and handle potential adversarial manipulations?

Monitoring and Detection

9. How are adversarial activities or anomalies detected during the model's deployment, and what response mechanisms are in place?

Model Updating

10. What is the process for updating the model to address newly identified adversarial attack methods or vulnerabilities?

Assurance Measures

11. What evidence or validations are available to demonstrate that the model's robustness measures are effectively implemented?

Access Control

12. How are access controls enforced to assure unauthorized querying or exploitation of the model (e.g., to mitigate model extraction)?

Defense against Inference Attacks

13. How does the system protect sensitive information from being inferred through adversarial techniques, such as inversion or querying?

Adversarial Awareness Training

14. What training or awareness programs are in place for developers and users regarding adversarial threats and defenses?

Impact Assessment

15. How is the potential impact of a successful adversarial attack on the confidentiality, integrity, or availability of the system assessed?

Model Transparency

16. How is transparency about the model's robustness to adversarial attacks balanced with protecting sensitive or proprietary information?

Robustness Validation Tools

17. What tools or frameworks are used to validate the model's robustness, and how often are these assessments conducted?

Industry Standards Compliance

18. How does the system align with industry standards or best practices for AI robustness and security?

Incident Response

19. What incident response plans are in place for handling adversarial attacks, and have these plans been tested?

Continuous Improvement

20. How is feedback from adversarial attack testing and real-world deployment used to continuously improve the system's robustness?

Performance Degradation

21. Can you describe any instances where the model's performance has inexplicably worsened over time, and what factors might have contributed to those changes?

Unexpected Behavior

22. What examples of unexpected behavior or unintended results has the model produced, and how has the training team investigated and addressed these issues?

Accuracy Fluctuations

23. How has the accuracy of the model been monitored over time, and what explanations have been identified for any significant, unexpected changes in its accuracy?

User Feedback on Model Decisions

24. How does the organization collect and respond to user feedback, particularly regarding spikes in problematic or incorrect decisions made by the model?

Bias Detection

25. What processes are in place to identify and address any tendencies for the model to return results that skew toward a particular direction or demographic, and how has bias been mitigated?

Security Events and Adversary Targeting

26. Has the organization experienced any attacks or security events that might suggest it is an active target? How were these events investigated for potential pathways to adversarial access or manipulation of training data?

Insider Threats

27. How does the organization monitor for unusual employee activities, such as a heightened interest in the intricacies of training data or the security measures protecting it? What actions are taken when such behavior is observed?

These questions are designed to encourage comprehensive responses, ensuring a thorough assessment of the AI system's adversarial robustness.

NOTES

1 The source for these questions resides with the GAO analysis of US and foreign government, industry, academic, and nonprofit organization documents, interviews, and forum discussions. Ariga, T., Persons, T., Sanford, S., June 2021, "Artificial Intelligence Accountability Framework, An Accountability Framework for Federal Agencies and Other Entities," US Government Accountability Office (GAO), GAO-21–519SP, https://www.gao.gov/products/gao-21-519sp.

2 Modffa, T., November 1, 2012, "Annex 2 – Information system security risk management activities (ITSG-33)," Canadian Centre for Cyber Security, www.cyber.gc.ca/en/guidance/annex-2-information-system-security-risk-management-activities-itsg-33#a72.

3 Lecuyer, M., Atlidakis, V., Geambasu, R., Hsu, D., and Jana, S., (n.d.)., "Certified Robustness to Adversarial Examples with Differential Privacy," Columbia University, Department of Computer Science, https://par.nsf.gov/servlets/purl/10112808.

4 Kotyan, S., and Vargas, D.V., April 14, 2022, "Adversarial Robustness Assessment: Why in Evaluation both L0 and L∞ Attacks are Necessary," *PLoS One*, vol. 17, no. 4, e0265723, doi: 10.1371/journal.pone.0265723. PMID: 35421125; PMCID: PMC9009601, www.ncbi.nlm.nih.gov/pmc/articles/PMC9009601.

Chapter 23

Auditing AI Systems

Continuous Monitoring

CONTINUOUS MONITORING OF AI SYSTEMS

Continuous monitoring refers to the ongoing process of collecting, analyzing, and evaluating data related to an artificial intelligence (AI) system's performance, safety, fairness, and compliance. This involves tracking key metrics, identifying deviations or anomalies, and ensuring the system operates as intended under dynamic conditions.

CONTINUOUS MONITORING OF DEPLOYED AI SYSTEMS

Continuous monitoring is necessary because AI systems often operate in complex, changing environments. New data, evolving user behavior, and external threats (like adversarial attacks) can impact system performance.

Monitoring is intended to ensure:

- Identification and mitigation of biases or ethical concerns.
- Maintenance of accuracy, fairness, and relevance.
- Early detection of failures or vulnerabilities.
- Compliance with regulatory and ethical standards.

CONTINUOUS MONITORING OF AI SYSTEMS: BASIC PROCEDURAL STEPS

The basic steps in the process of auditing an organization's policies and procedures and the ability to continuously monitor its AI systems include:

- Policy Review: Analyze existing organizational policies to ensure alignment with AI-specific risks, including fairness, accountability, and security.

DOI: 10.1201/9781003646501-28

- Capability Assessment: Evaluate the technical and operational capabi lities of the organization to support continuous monitoring, such as tools for data logging, anomaly detection, and feedback loops.
- Framework Development: Create or adopt frameworks for systematic monitoring, including defining key metrics, thresholds, and alert mechanisms.
- Implementation of Monitoring Tools: Integrate automated monitoring systems to track performance, compliance, and ethical considerations.
- Training and Awareness: Ensure staff are trained to manage and interpret monitoring outputs and address identified issues.
- Regular Audits: Conduct periodic reviews to assess the effectiveness of monitoring systems and refine policies and procedures as needed.

GAO AI ACCOUNTABILITY FRAMEWORK – AUDITING AI CONTINUOUS MONITORING

The US Government Accountability Office (GAO) developed the Artificial Intelligence Accountability Framework in part to help organizations ensure the reliability and relevance of AI systems over time. In this framework, the GAO created a set of specific audit questions aimed at AI Continuous Monitoring.

The objectives of these questions are to:

- Develop plans for continuous or routine monitoring of the AI system to ensure it performs as intended.
- Establish the range of data and model drift that is acceptable to ensure the AI system produces the desired results.
- Document results of monitoring activities and any corrective actions taken to promote traceability and transparency.
- Assess the utility of the AI system to ensure its relevance to the current context.
- Identify conditions, if any, under which the AI system may be scaled or expanded beyond its current use.

Auditing AI Continuous Monitoring – GAO AI Accountability Framework Questions

The following is a list of audit questions designed to assess an organization's ability to meet the above AI Continuous Monitoring objectives.

1. What plans has the organization developed to monitor the AI system?
2. To what extent do the plans describe processes and procedures to continuously monitor the AI system?

3. What is the established frequency for monitoring the AI system?
4. To what extent is the frequency feasible and appropriate for effectively managing system performance?
5. To what extent has the organization established an acceptable range for the data and model drift?
6. To what extent was the acceptable range for the data and model drift established based on a risk assessment, and is it appropriate for its use case?
7. What mechanisms have been developed to detect data and model drift?
8. To what extent do the monitoring and tuning activities track performance and avoid undesired consequences?
9. To what extent did the organization document the frequency and rationale for updating the AI system?
10. To what extent did the organization document the results of monitoring activities and corrective actions?
11. How does the organization determine the ongoing utility of the AI system?
12. To what extent is the AI system still needed to achieve goals and objectives?
13. To what extent has the organization identified metrics that will help it determine whether and when to retire the AI system?
14. To what extent does the organization have established procedures for retiring the AI system, if it is no longer needed?
15. What assessments and evaluations have been completed to determine if the AI system can be applied to address otheEr issues/problems?
16. To what extent did these assessments and/or evaluations identify conditions under which such applications can or cannot take place?
17. How did the organization use these assessments and/or evaluations to determine if the system can be scaled up?
18. To what extent can the AI system be applied to different use cases or problems?[1]

NOTE

1 The source for these questions resides with the GAO analysis of US and foreign government, industry, academic, and nonprofit organization documents, interviews, and forum discussions. Ariga, T., Persons, T., and Sanford, S., June 2021, "Artificial Intelligence Accountability Framework, An Accountability Framework for Federal Agencies and Other Entities," US Government Accountability Office (GAO), GAO-21-519SP, www.gao.gov/assets/gao-21-519sp.pdf.

Chapter 24

Data Collection and Operational Consistency

DATA COLLECTION AND OPERATIONAL CONSISTENCY

Auditing artificial intelligence (AI) for consistency of data and operations involves a thorough review and evaluation of the AI system's processes and data handling to ensure accuracy, reliability, and ethical compliance. This audit focuses on identifying potential biases, verifying the consistency and quality of data, and ensuring that the AI model's operations align with established standards and best practices.

Objectives:

1. Identify and Reduce Bias: Ensure that the AI system does not produce biased or discriminatory outcomes by evaluating data sources and model predictions for fairness.
2. Assure Data Consistency: Verify that data used throughout the AI lifecycle (training, validation, testing) is consistent in quality, format, and processing.
3. Ensure Operational Consistency: Confirm that the AI system operates reliably and consistently, following documented procedures and protocols.

AUDITING AI – DATA COLLECTION AND OPERATIONAL CONSISTENCY

Data Collection

1. What are the sources of data used for training, validation, and testing of the AI model?
2. How is the quality of the data assessed and ensured before use?

DOI: 10.1201/9781003646501-29

- **Bias Detection and Mitigation**
1. What methods are used to identify potential biases in the data and model predictions?
2. What steps have been taken to mitigate identified biases in the AI system?

- **Data Preprocessing and Transformation**
1. What preprocessing steps are applied to the data, and how are these steps documented and consistently applied?
2. How are missing values, outliers, and anomalies handled in the data?

- **Model Training and Validation**
1. What performance metrics are used to evaluate the AI model, and how do they ensure fairness across different demographic groups?
2. How is the model validated to ensure it performs well on diverse and representative datasets?

Operational Consistency

1. What procedures are in place to ensure the AI model operates reliably and consistently across different environments?
2. How are updates and changes to the AI model documented and communicated?

- **Transparency and Explainability**
1. How are the AI model's decisions and predictions explained and documented?
2. What methods are used to provide transparency in the model's decision-making processes?

- **Data Security and Privacy**
1. What measures are in place to protect the data used by the AI system from unauthorized access and breaches?
2. How is compliance with data protection regulations ensured?

- **User and Stakeholder Feedback**
1. How is feedback from users and stakeholders collected and integrated into the AI system's development and improvement?
2. Are there mechanisms in place to address concerns and issues raised by users regarding the AI system?

- **Continuous Monitoring and Improvement**
1. What procedures are in place for continuous monitoring of the AI system's performance and outcomes?

2. How is the AI model regularly updated and improved to maintain its effectiveness and relevance?

- **Compliance and Ethical Considerations**

1. How does the AI system comply with relevant regulations, standards, and ethical guidelines?
2. Are there documented procedures for handling ethical concerns and ensuring responsible use of the AI system?

DATA MANAGEMENT

Data used to train, test, and validate AI systems should be of sufficient quality and appropriate for the intended purpose to ensure the system produces consistent and accurate results.[1] Management should provide reasonable assurance of the quality, reliability, and representativeness of the data included in the system, from its development stage to system operation.[2]

Data may be separated into two distinct categories. These categories follow.

DATA USED FOR MODEL DEVELOPMENT

Data used for model development refers to datasets employed during the creation and refinement of an AI model. This includes:

- Training Data: Data used to train the model, enabling it to learn patterns and relationships.
- Validation Data: Data used to fine-tune model parameters and prevent overfitting during training.
- Test Data: Data used to evaluate the model's performance and generalization on unseen examples.

These datasets are typically curated to represent the problem domain and ensure the model's accuracy, reliability, and fairness.

Data Used for System Operations

Data used for system operations refer to real-world input data processed by the AI system after deployment. This includes:

- Operational Data: Data fed into the system for predictions or decision-making in live environments.

- Monitoring Data: Data collected to monitor the system's performance, identify anomalies, or ensure compliance.
- Feedback Data: Data from user interactions or system outputs, which may inform future updates or retraining.

Operational data should align with the characteristics of the development data to maintain system accuracy and robustness.

Auditors should consider the following three sources of evidence to assess data used for AI models, including documentation on how data

1. Have been collected.
2. Prepared, labeled, and maintained.
3. Are monitored continually.

Organizations should also ensure that data used to develop the models in the AI models are reliable because data reliability affects the accuracy of model predictions.

In an audit environment:

- **Reliability** of data means that data are applicable for audit purposes and are sufficiently complete and accurate.
- **Applicability for audit purposes** refers to whether the data, as collected, are valid measures of the underlying concepts being addressed in the objectives.
- **Completeness** refers to the extent to which relevant data records and fields are present and sufficiently populated.
- **Accuracy** refers to the extent to which recorded data reflect the actual underlying information.
- **Consistency** refers to whether data are sufficiently clear and well-defined to yield comparable results in similar analyses.[3]

The organization should implement internal control procedures to reasonably ensure that:

1. Data are inputted in a controlled manner.
2. Data inputted into the application are complete, accurate, and valid.
3. Any incorrect information is identified, rejected, and corrected for subsequent processing.
4. The confidentiality of the data is adequately protected.[4]

Inadequate controls can result in incomplete, inaccurate, and/or invalid records in the application data or unauthorized disclosure of application data.

Organizations should also assess the extent to which the data accurately and verifiably represent constituent populations served by the AI system. AI models trained on data that are not representative of the target population may produce biased results, in that the model performs well for the characteristics that are well-represented and poorly for the characteristics that are underrepresented.

Another key data quality is its representativeness. Issues with representativeness may arise in multiple ways: Characteristics that are significant for model development are not sufficiently represented in the data (e.g. because of sampling or coverage bias). In such cases, the model will show poor performance in both development and production.

The model may perform well for characteristics that are represented well in the data, but it may underperform for characteristics that are underrepresented (e.g., a facial recognition algorithm trained on data with pictures of a certain ethnicity will perform well for phenotypes sufficiently represented and badly for phenotypes that are underrepresented).

There are new issues concerning data that are specific to ML modeling. One of the more well-known issues is the lack of separation between training and testing/validation data. When a part of the data is used both for shaping the model (during the training phase) and verifying the performance of the model (during the testing or validation phase), the performance metrics of the model will be inflated. Model performance would then drop in production.[5]

Organizations should be aware of the risks associated with data used in AI models that include:

- Usage of unreliable, biased, and non-representative raw data.
- Biasing the data during data transformation.
- Insufficient separation between training and testing/validation data.
- Poor generalizability of the model on new data.
- Poisoning – adversarial introduction of bad-quality data.
- Target leakage.[6]

GAO AI Accountability Framework – Auditing Data Used for Model Development

The US Government Accountability Office (GAO), in developing the Artificial Intelligence Accountability Framework, created a set of specific audit questions aimed at the AI data framework.

The objective of these questions is to:

- Document sources and origins of data used to develop the models underpinning the AI system.
- Assess

- The reliability of data used to develop the models.
- Attributes used to categorize data (categorization).
- Data variables used in the AI component models.
- The use of synthetic, imputed, and/or augmented data.[7]
- Interconnectivities and dependencies of data streams that operationalize the AI system.
- The reliability, quality, and representativeness of all the data used in the system's operation, including any potential biases, inequities, and other societal concerns associated with the AI system's data.
- Data security and privacy for the AI system.

AUDITING AI DATA – GAO AI ACCOUNTABILITY FRAMEWORK QUESTIONS

The following is a list of audit questions designed to assess an organization's ability to meet the above AI data framework objectives.

1. How has the organization documented the AI system's data provenance, including sources, origins, transformations, augmentations, labels, dependencies, constraints, and metadata?
2. What processes exist for data generation, acquisition/collection, ingestion, staging/storage, transformations, security, maintenance, and dissemination?
3. To what extent are the data appropriate for the intended purpose?
4. To what extent are data used to develop the AI system accurate, complete, and valid?
5. To what extent do the data represent constituent populations served by the AI system?
6. How does the organization ensure that the data collected are adequate, relevant, and not excessive with the intended purpose?
7. What corrective actions has the organization taken to enhance the quality, accuracy, reliability, and representativeness of the data?
8. What attributes are used to categorize data?
9. To what extent are the data attributes accurate, complete, and valid?
10. What is the method for segregating data into training, validation, and testing sets?
11. To what extent is the training, validation, and testing data representative of the operational environment?
12. What assumptions, if any, were made about the operational environment?
13. What is the variable selection and evaluation process?
14. How were sensitive variables (e.g., demographic and socioeconomic categories) that may be subject to regulatory compliance specifically selected or not selected for modeling purposes?

15. What is the organization's rationale for using synthetic, imputed, and/or augmented data?
16. How are synthetic (artificially generated data for training or testing AI models), imputed data (filling in missing data for model training), and/or augmented data (creating new data from existing data to improve model performance) generated, maintained, and integrated?
17. What assumptions, if any, were made in the process of generating synthetic imputed data and/or augmented data?
18. What goals have been set by the organization to evaluate the model's performance, thereby reducing the need to augment data?
19. How will the organization be able to determine if the model's performance has reached diminishing returns, thus identifying the point at which data augmentation is no longer necessarys?
20. What evidence exists that the organization has tested the robustness of training data on real-world data?
21. To what extent has the organization identified and mitigated potential bias – statistical, contextual, and historical – in the data?
22. How has the organization identified and mitigated potential impacts of bias in the data, including inequitable or discriminatory outcomes?
23. What assessments has the organization conducted on data security and privacy impacts associated with the AI system?
24. How does the organization identify, assess, and mitigate data security and privacy risks associated with the AI system?[8]

DATA MANAGEMENT – GENERAL AUDIT QUESTIONS

1. How does your organization ensure the accuracy and completeness of the data inputs used by the AI system?
2. What criteria are used to assess the quality and relevance of the data inputs for the AI system?
3. Can you describe the processes in place to regularly validate and update the data used by the AI system?
4. How does your organization handle and rectify data discrepancies or inaccuracies that may affect the AI system's outputs?
5. What measures are implemented to ensure the privacy and security of the data inputs used by the AI system?
6. How are data sources evaluated and selected for use in the AI system?
7. What internal controls are established to monitor and mitigate the risks associated with data quality issues in the AI system?
8. Can you explain the procedures for ensuring that the data used by the AI system complies with relevant data protection regulations, such as GDPR?
9. How does your organization handle sensitive or personally identifiable information within the data inputs for the AI system?

10. What processes are in place to document and review the lineage of data inputs, including their origin, transformations, and usage within the AI system?

NOTES

1 Machine learning systems learn from data, known as the training set, in order to perform a task. A testing set is a subset of the data used to test the trained model. The testing set should be large enough to yield statistically relevant results and be representative of the data set as a whole.
2 Ariga, T., Persons, T., and Sanford, S., June 2021, "Artificial Intelligence Accountability Framework, An Accountability Framework for Federal Agencies and Other Entities," US Government Accountability Office (GAO), GAO-21–519SP, https://www.gao.gov/products/gao-21-519sp.
3 Moessbauer, S., Fejfar, M., Lauber, K., and Adkins, N., et al., December 16, 2019, "Assessing Data Reliability," US Government Accountability Office, GAO-20–283G, https://www.gao.gov/products/gao-20-283g
4 Simpson, D., Burkart, N., Perkins, R., and Bradley, C., et al., September 5, 2024, "Federal Information System Controls Audit Manual," Government Accountability Office, GAO-24–107026, https://www.gao.gov/products/gao-24-107026.
5 Supreme Audit Institutions of Finland, Germany, the Netherlands, Norway, and the UK, April 7, 2023, "Auditing machine learning algorithms," www.auditingalgorithms.net.
6 **Target leakage** refers to a situation where information from the target variable (the outcome the model is trying to predict) is inadvertently included in the training data in a way that would not be available in real-world prediction scenarios. This causes the model to learn patterns that are not valid during deployment, leading to overly optimistic performance during training and poor generalization of unseen data.
7 **Synthetic Data:** Artificially generated data that mimics real-world data distributions, often created using algorithms or models (e.g., GANs) to address privacy concerns, enhance data diversity, or supplement limited datasets.
Imputed Data: Data that replaces missing or incomplete values with estimated or calculated values based on patterns or statistical techniques to ensure consistency and usability.
Augmented Data: Data that has been modified or expanded using transformations (e.g., rotations, scaling, or noise addition) to improve model robustness, particularly in scenarios like image or text processing.
8 The source for these questions resides with the GAO analysis of U.S. and foreign government, industry, academic, and nonprofit organization documents, interviews, and forum discussions. Ariga, T., Persons, T., Sanford, S., June2021,"ArtificialIntelligenceAccountabilityFramework,AnAccountability Framework for Federal Agencies and Other Entities," U.S. Government Accountability Office (GAO), GAO-21-519SP, www.gao.gov/assets/gao-21-519sp.pdf. (Note: Questions #16, 17, 18, 19, and 20, where developed by the author).

Appendix I

Auditing AI: General Audit Overview and Questions

AUDIT OBJECTIVE

Evaluate the AI system's design, implementation, and effectiveness within the organization. This includes assessing whether the AI system operates following its intended purpose, aligns with organizational policies, complies with relevant legal and regulatory requirements, and effectively mitigates associated risks.

AUDIT SCOPE

The following areas of AI will be examined using the general AI audit questions in this appendix.

- Change Management: Assess the processes for managing changes to the AI system, including updates, patches, and modifications, to ensure they are controlled and do not negatively impact system performance.
- Compliance and Regulatory Requirements: Ensure that the AI system complies with applicable laws, regulations, and industry standards, including data protection, privacy, and AI-specific regulations.
- Data Management: Assess the data inputs used by the AI system, including the quality, accuracy, completeness, and relevance of data, as well as data privacy and security measures.
- Documentation and Reporting: Evaluate the adequacy and completeness of documentation related to the AI system, including policies, procedures, user guides, and audit trails, as well as the effectiveness of reporting mechanisms.

- Ethical Considerations: Evaluate the ethical implications of AI usage, including potential biases, transparency, accountability, and adherence to ethical standards.
- Governance and Oversight: Review the governance framework for AI, including the roles, responsibilities, and accountability structures in place for overseeing AI-related activities.
- Model Development and Validation: Examine the process of AI model development, including model selection, training, testing, validation, and monitoring for ongoing performance and fairness.
- Risk Management: Identify and assess the risks associated with the AI system, including operational, reputational, and strategic risks, and evaluate the effectiveness of the risk management practices in place.
- Security: Review the security measures implemented to protect the AI system from unauthorized access, cyber threats, and other vulnerabilities.

AUDIT APPROACH

Pre-Audit Planning

- **Understanding the AI System:** Obtain a detailed understanding of the AI system, including its architecture, purpose, data sources, and integration with other systems.
- **Stakeholder Identification:** Identify key stakeholders involved in the AI system, including IT, data scientists, legal, compliance, and business units.
- **Preliminary Risk Assessment:** Conduct a preliminary risk assessment to identify high-risk areas within the AI system that require more detailed attention.
- **Document Review:** Collect and review relevant documentation, including governance policies, data management procedures, model development guidelines, ethical guidelines, compliance records, and security protocols.

Change Management Review

- Evaluate the change management process, including the handling of updates, patches, and system modifications.
- Review documentation and testing procedures to ensure changes do not adversely affect system performance.

Compliance and Regulatory Review

- Verify compliance with relevant legal and regulatory requirements, including data protection laws, AI-specific regulations, and industry standards.
- Review audit trails and evidence of compliance.

Data Management Review

- Analyze the quality, accuracy, and completeness of data inputs used by the AI system.
- Evaluate data privacy measures and verify compliance with data protection regulations.

Documentation and Reporting Verification

- Assess the adequacy and completeness of AI-related documentation.
- Verify the effectiveness of reporting mechanisms and the clarity of communication to stakeholders.

Ethical Considerations Assessment

- Assess the ethical implications of AI deployment, including bias detection, transparency measures, and accountability structures.
- Review adherence to ethical standards and industry best practices.

Governance and Oversight Evaluation

- Interview key personnel to assess the effectiveness of the governance structure and oversight mechanisms.
- Review meeting minutes, decision logs, and accountability matrices to ensure roles and responsibilities are clearly defined.

Model Development and Validation Testing

- Review the model development process, including data preprocessing, model selection, and training methods.
- Test the validation process by reviewing model performance metrics and testing the model against new data to evaluate accuracy and fairness.

Risk Management Analysis

- Identify potential risks associated with the AI system, including operational, reputational, and strategic risks.
- Evaluate the effectiveness of risk management strategies and controls in place.

Security Testing

- Perform security assessments to identify vulnerabilities in the AI system.
- Review access controls, encryption methods, and incident response plans.

Audit Reporting

- Findings and Recommendations: Document audit findings, highlighting areas of non-compliance, risks, and inefficiencies. Provide clear, actionable recommendations for improvement.

- Management Response: Obtain and document management's response to each finding and recommendation.
- Final Audit Report: Prepare a comprehensive audit report summarizing the audit objective, scope, approach, findings, and management responses. Include an executive summary for senior management.

Follow-Up and Monitoring

- Action Plan Review: Ensure management develops an action plan to address audit findings and recommendations.
- Follow-Up Audits: Schedule follow-up audits to verify the implementation and effectiveness of corrective actions.
- Ongoing Monitoring: Establish mechanisms for continuous monitoring of AI system performance, compliance, and risk management.

AUDIT TEAM

The audit team ideally will consist of professionals with backgrounds that include data science, legal and compliance, cybersecurity, internal controls, and domain-specific knowledge, to conduct the audit effectively.

General AI Audit Questions

1. Has the organization identified the bounds of responsible/intended uses for the AI tool?
2. How are data quality and integrity maintained?
3. How are privacy and data protection laws addressed and complied with?
4. How are stakeholders involved in the AI system's deployment and use? Are there any biases in the data, and how are they addressed?
5. How does the AI system handle edge cases or anomalies?
6. How transparent is the AI system in terms of its decision-making process?
7. How will compromise or misuse of the AI tool be identified?
8. Is there any human oversight or intervention in the AI system's decisions?
9. Is there a procedure for when automated decisions or activities of the AI tool will require human approval or to be marked out for human oversight?
10. What data does the AI system use, and what are the sources of this data?
11. Once deployed, what degree of direct human control is needed for the AI tool?
12. What ethical guidelines govern the operation of the AI system?
13. What is the error rate of the AI system, and how are errors handled?

14. What is the primary function of the AI system, and what decisions does it inform?
15. What measures are in place to ensure the AI system's outputs are accurate and reliable?
16. What processes are in place for continuous monitoring and updating of the AI system?

Appendix 2
Auditing AI Algorithms

AUDIT OBJECTIVE

To ensure that the AI system is trustworthy and that its algorithms support responsible and equitable outcomes.

AUDIT SCOPE

The audit will evaluate the AI's design, development, and deployment processes to ensure that the algorithms perform correctly, fairly, and ethically within their intended operational environments. This includes assessing the algorithms for accuracy, transparency, accountability, non-discrimination, and compliance with relevant regulations and standards.

The audit will also seek to identify any risks or issues related to data handling, model training, output interpretation, and decision-making processes, and to verify that appropriate governance, oversight, and control mechanisms are in place.

AUDIT APPROACH

Auditing an AI's algorithms is a multifaceted process that involves evaluating the design, implementation, and impact of the algorithms to ensure they function as intended, and to verify that they comply with legal, ethical, and operational standards.

The audit will evaluate specific areas related to the functionality of the AI software. These areas include but are not limited to:

- Algorithmic Decision-Making
- Data Management
- Ongoing Performance Monitoring.

This will inform the development of detailed audit procedures and criteria against which the AI's algorithms will be evaluated.

AUDIT TEAM

The audit team ideally will consist of professionals with backgrounds that include data science, legal, compliance, cybersecurity, internal controls, and domain-specific knowledge, to conduct the audit effectively.

Steps to Auditing AI Algorithms

Step 1 Understand the AI System

Gain a thorough understanding of the AI system's purpose, capabilities, and operational context. This includes understanding the data it uses, the models it employs, the outputs it generates, and the decisions it informs.

Step 2 Define the Audit Scope

Determine the specific aspects of the AI system to be audited. This could include data management, model training, output analysis, or the decision-making process.

Step 3 Assess Data Management

Evaluate the data collection, storage, and processing practices to ensure data quality, integrity, and compliance with data protection regulations.

Step 4 Evaluate Model Training

Inspect the algorithms for biases, examine the training process, and validate the training data to ensure the model is trained appropriately for its intended use.

Step 5 Test the AI Model

Conduct rigorous testing to assess the model's performance, accuracy, robustness, and how it handles edge cases.

Step 6 Review Outcomes and Decision Processes

Analyze the decisions made by the AI to identify any potential biases or unethical outcomes.

Step 7 Compliance and Ethical Considerations

Ensure that the AI system adheres to relevant laws, regulations, and ethical guidelines, including transparency, accountability, and fairness.

Step 8 Documentation and Transparency

Review the documentation for clarity, completeness, and transparency. The AI system's workings should be understandable to the intended audience.

Step 9 Stakeholder Engagement

Engage with stakeholders to understand their perceptions and concerns about the AI system.

Step 10 Continuous Monitoring

Establish procedures for ongoing monitoring and auditing of the AI system to ensure sustained compliance and performance.

AUDITING AN AI'S ALGORITHMS

Audit Program Questions

Auditing AI Algorithms – General Questions

1. Are legal documents drafted that allow for a review of third-party propriety algorithms without violating intellectual property (IP) rights?
2. Are the AI tool's Model Cards examined for purpose, performance metrics, training data, and limitations?
3. Are the outputs of the algorithm intelligible to those personnel making decisions based on the algorithm?
4. Could the algorithm cause harm to organization personnel and/or clients?
5. Does the use of the algorithm raise ethical, legal, or social concerns that through its use will place the organization at risk?
6. Has the use of the algorithm been shown to improve an organization's strategic objectives and outcomes?
7. How are the algorithms and models trained, validated, and tested?
8. How good were the data used to train the algorithm?
9. How well does the algorithm perform?
10. How will this algorithm fit into and complement current workflows?
11. Is the algorithm transferable to new organizational settings or uses?
12. Were there sufficient data to train the algorithm?
13. What algorithms and models does the AI system employ?
14. What is the purpose and context of the algorithm? To what extent were the data accurate and free of bias? Were the data standardized?
15. Were there sufficient data to train the algorithm?

Auditing AI Algorithms – Data Sources Model Training and Deployment

Data Sources used in AI model training and development refer to the datasets and information repositories from which data is obtained to train, validate, and test AI models. These sources can be diverse and include structured data (databases, spreadsheets), unstructured data (text, images, videos), and real-time data (sensor feeds, APIs).

Key Types of Data Sources:

1. Structured Data: Organized data in tabular format, such as databases and spreadsheets.
2. Unstructured Data: Data that does not have a predefined structure, such as text documents, images, and videos.
3. Semi-Structured Data: Data with some organizational properties but not as rigid as structured data, such as JSON and XML files.
4. Real-Time Data: Data collected in real-time from sensors, IoT devices, or streaming APIs.
5. Historical Data: Archived data collected over time, often used for time-series analysis and forecasting.

Audit of the AI Model and AI Tool – Documentation Examination

In the course of an audit of the AI model and overall, the AI tool, the auditor should request the following documentation. Cross-referencing this documentation with the logs of the data flows within the AI tool (see separate audit program section on data flow logs), will provide a clearer picture of how and why the algorithm produced the response it did to the user's prompt.

Data Source Documentation

- Origin: Source of the data, including provider details and acquisition method.
- Description: Detailed description of the data, including format, structure, and content.
- Quality: Information on data quality, including any known issues, missing values, and anomalies.
- Processing: Preprocessing steps applied to the data, such as cleaning, normalization, and transformation.

Data Usage Documentation

- Purpose: The purpose of using the data in the AI model, including training, validation, and testing.
- Volume: Size and scope of the data used in different phases of model development.
- Access Controls: Security measures and access controls in place to protect the data.

Data Provenance

- Lineage: Traceability of the data from its source to its use in the model, ensuring transparency in data handling.
- Metadata: Additional information about the data, such as creation date, authorship, and modifications.

Role of Data Flow Logs in the AI Tool

Data Flow Logs are records of how data moves through an AI system, from ingestion to processing, storage, and output.

Data flow logs play a crucial role in understanding and auditing AI tools for several reasons:

1. Traceability: Logs provide a detailed account of data movement, enabling auditors to trace the data's journey through the system.
2. Decision Tracking: By examining data flow logs, auditors can understand how specific data points influence model decisions, ensuring transparency in decision-making processes.
3. Error Detection: Logs help identify errors and inconsistencies in data handling, allowing for timely resolution and ensuring data integrity.
4. Compliance: Maintaining detailed logs helps ensure compliance with data protection regulations and standards, as they provide evidence of proper data management practices.

EXAMINING DATA FLOW LOGS

When auditing an AI tool, specific aspects of data flow logs should be examined:

- Data Ingestion:
  1. What data sources are being ingested into the system?
  2. Are there any inconsistencies or errors in data ingestion?

- Data Processing:
  1. What transformations and preprocessing steps are applied to the data?
  2. Are these steps documented and consistent with the data processing pipeline?
- Data Storage:
  1. How is the data stored, and what security measures are in place?
  2. Are there any discrepancies between documented storage procedures and actual practices?
- Data Access:
  1. Who has access to the data at various stages of its lifecycle?
  2. Are access controls properly implemented and logged?
- Data Outputs:
  1. How is the processed data used in the model's decision-making?
  2. Are the outputs consistent with the expected results based on the input data?

The auditor should also review the following audit program sections for additional insight and audit questions related to the assessment of the AI algorithmic decision-making process:

- Auditing AI – AI Model Cards
- Auditing AI – AI System Cards
- Auditing AI – AI Data Cards

These individual audit program questions and additional learning materials will be found in the main body of the AI audit program (see Chapter 16 for further detailed discussion of these AI cards).

AUDITING AI ALGORITHMS – ETHICAL, LEGAL, AND SOCIAL ISSUES

1. How were consent issues handled in collecting data used for the algorithm training and validation?
2. Who owns, or has stewardship of, the data and determines how it is to be used in the training and testing of algorithms?
3. How are data confidentiality and privacy ensured when data is stored (in the cloud) used and shared across different platforms?

4. How much responsibility for care should employees be expected to assume when using algorithms they cannot control or explain?
5. Who carries liability resulting from a faulty or misapplied algorithm (developers who trained and tested the algorithm, vendors who integrated the algorithm into electronic medical records or imaging software, or employees using the algorithm to make decisions)?
6. Who takes responsibility for postimplementation monitoring of the safety and efficacy of an algorithm throughout its lifecycle, and determines when an algorithm needs updating, retraining, or even withdrawal because of emerging inaccuracies?
7. Will the majority of organization personnel responsible for managing and using the AI tool be literate enough to understand how, when, and by whom machine-learning algorithms are safe and effective to use?
8. How equitable and inclusive are the algorithms? Will a digital divide create a liability or risk for the organization?
9. Who within the organization might have conflicts of interest in developing, disseminating, using, or advocating a particular algorithm?
10. For internally developed AI algorithms, who owns the IP of an algorithm; who owns the patent rights, and who and what factors determine whether an algorithm can be commercialized for profit?

AUDITING AI ALGORITHMS – MODEL INSPECTION SOFTWARE

Model Inspection Software refers to tools and applications designed to analyze and evaluate AI models. These tools provide detailed insights into the model's behavior, performance, and decision-making processes. Model Inspection Software typically includes features for visualizing model predictions, identifying biases, and validating the model's accuracy and robustness.

Key Components of Model Inspection Software:

- Model Visualization: Tools to visualize the model's architecture, decision boundaries, and feature importance.
- Performance Metrics: Detailed performance evaluation, including accuracy, precision, recall, F1-score, and other relevant metrics.
- Bias Detection: Identification of potential biases in the model's predictions and underlying data.
- Explainability: Methods to explain the model's predictions, such as feature attribution and decision trees.
- Robustness Testing: Evaluation of the model's performance under various conditions and edge cases.
- Compliance Checks: Ensuring the model adheres to regulatory and ethical standards.

Usefulness in AI Tool Audits

Using Model Inspection Software during an audit can significantly enhance the evaluation process by providing in-depth analysis and understanding of the AI model. They are most appropriately used to assess areas related to:

- Model Performance: Verifying the accuracy, precision, and robustness of the AI model.
- Bias and Fairness: Identifying and mitigating potential biases in the model's predictions.
- Transparency and Explainability: Understanding the model's decision-making processes and ensuring they can be explained.

Auditing the AI Tool Using the Tool's AI Data Card

- Model Performance:
  1. What are the key performance metrics of the AI model, and how do they compare with the expected benchmarks?
  2. How does the model perform across different datasets and scenarios?
- Bias Detection:
  1. Are there any identified biases in the model's predictions?
  2. How does the model's performance vary across different demographic groups?
- Model Explainability:
  1. How are the model's predictions explained, and what methods are used for feature attribution?
  2. Are the decision-making processes of the model transparent and understandable?
- Robustness Testing:
  1. How does the model perform under various conditions, including edge cases and adversarial scenarios?
  2. What steps have been taken to ensure the model's robustness and reliability?
- Compliance and Ethical Standards:
  1. How does the model comply with relevant regulations and ethical guidelines?
  2. Are there documented procedures for handling ethical concerns and potential biases?
- Visualization Tools:
  1. What visualization tools are available to analyze the model's architecture and decision boundaries?
  2. How do these visualizations help in understanding the model's behavior?

- Feature Importance:
  1. What features are most important in the model's predictions, and how is this determined?
  2. Are there any features that disproportionately influence the model's decisions?
- Data Quality:
  1. How is the quality of the data used to train the model assessed and ensured?
  2. Are there any issues related to data quality that impact the model's performance?
- User and Stakeholder Feedback:
  1. How is feedback from users and stakeholders collected and integrated into the model's evaluation and improvement process?
- Continuous Monitoring and Updates:
  1. What procedures are in place for continuous monitoring and updating of the model?
  2. How often is the model reviewed and updated to ensure ongoing effectiveness and compliance?

Appendix 3

Explanation of the Noise Vector Generation Function

The statement "is the noise vector generation function creating appropriate input data for the GAN model?" ensures that the random noise vectors fed into the Generative Adversarial Network (GAN) model are suitable and correctly formatted for the model to generate meaningful outputs.

WHAT IS A NOISE VECTOR IN GANS?

A noise vector is a random vector sampled from a specific distribution, typically a Gaussian (normal) distribution or a uniform distribution. This vector is used as input to the generator part of the GAN. The generator then transforms this noise vector into a synthetic data instance, such as an image.

IMPORTANCE OF THE NOISE VECTOR

The quality and distribution of the noise vectors are crucial because they affect the generator's ability to produce diverse and realistic outputs. The noise vector essentially seeds the generator, providing the starting point for the synthetic data generation process.

DETERMINING APPROPRIATENESS OF THE NOISE VECTOR GENERATION

To determine if the noise vector generation function is creating appropriate input data for the GAN model, the audit procedure is as follows:

Distribution:

- Expected Distribution: Verify that the noise vectors follow the expected distribution (e.g., Gaussian or uniform). This is typically defined by the GAN architecture.
- Parameter Validation: Ensure that the parameters of the distribution (mean, standard deviation for Gaussian; range for uniform) match the requirements of the model.

Dimensionality:

- Correct Dimensionality: Check that the noise vectors have the correct dimensionality. For example, if the GAN model expects a 100-dimensional noise vector, the function should generate vectors of this dimension.

Range and Scaling:

- Value Range: Ensure that the values within the noise vector fall within the expected range. For instance, if the model expects values between –1 and 1, the generated noise vectors should adhere to this range.
- Scaling: Verify if any scaling or normalization is needed and ensure it is applied correctly.

Implementation:

- Code Review: Examine the implementation of the noise vector generation function to ensure it is correctly producing the vectors as intended.
- Unit Testing: Implement unit tests to generate noise vectors and check their properties (distribution, dimensionality, range).

Example: Generating a Noise Vector

Here's a brief example of generating a noise vector for a GAN:

```
import numpy as np
def generate_noise_vector(dimensions, distribution='gaussian'):
if distribution == 'gaussian':
return np.random.randn(1, dimensions) # Standard normal distribution
elif distribution == 'uniform':
return np.random.uniform(-1, 1, (1, dimensions)) # Uniform distri-
bution between -1 and 1
else:
raise ValueError("Unsupported distribution type")
# Example usage
noise_vector = generate_noise_vector(100, 'gaussian')
print(noise_vector)
```

Ensuring that the noise vector generation function is creating appropriate input data for the GAN model involves verifying the correct distribution, dimensionality, value range, and implementation of the noise vectors.

Properly generated noise vectors enable the GAN to effectively learn and produce high-quality synthetic data.

Appendix 4

Identifying and Retrieving an AI Model's F1 Score Data: Audit Procedure

The F1 score is a metric used to evaluate the performance of a binary classification model. It is the harmonic mean of precision and recall, providing a single measure that balances both aspects. The F1 score is particularly useful when the dataset has an imbalanced class distribution.

The auditor can find and extract the data necessary to calculate or verify an AI model's F1 score from several sources within the machine learning pipeline.

Here are the key places and steps they should follow:

1. Model Output and Predictions
    A. Predicted Labels (Model Predictions)
        Location: These are typically found in the output files or logs generated by the model during inference/testing. In many machine learning platforms, these might be stored in structured databases, flat files (e.g., CSV, JSON), or specialized data storage systems.
        Extraction: Use scripts or data query tools to extract the predicted labels for the test dataset. Ensure the data is correctly matched with the corresponding test samples.
2. Ground Truth Data
    B. True Labels (Ground Truth)
        Location: These are found in the ground truth dataset used for validation. They might be stored in databases, annotation files, or in the same files as the test dataset with true labels included.
        Extraction: Access the source where the true labels are stored and extract them. Ensure the labels are correctly aligned with the test samples used for predictions.

3. Combining Predictions and Ground
   C. Data Alignment
      Ensure that the predicted labels and true labels are correctly aligned. Each prediction should correspond to its respective true label.
      Verification: Check for any mismatches or misalignments that could affect the accuracy of the F1 score calculation.
4. Calculate the Precision, Recall, and F1 Score
   D. To calculate an AI model's F1 score use the following formula.
      Precision: The ratio of true positive predictions to the total predicted positives (i.e., how many of the predicted positive cases were correct).

$$\text{Precision} = \frac{\text{True Positives (TP)}}{\text{True Positives (TP)} + \text{False Positives (FP)}}$$

Recall: The ratio of true positive predictions to the total actual positives (i.e., how many of the actual positive cases were correctly identified)

$$\text{Recall} = \frac{\text{True Positives (TP)}}{\text{True Positives (TP)} + \text{False Negatives (FN)}}$$

The F1 score combines these two metrics using the following formula:

$$\text{F1 Score} = 2 \times \frac{\text{Precision} \times \text{Recall}}{\text{Precision} + \text{Recall}}$$

A high F1 score indicates a good balance between precision and recall, meaning the model performs well in identifying positive cases without many false positives or false negatives.
5. Documentation and Verification
   E. Accessing Model Evaluation Reports
      Location: Model evaluation reports often contain detailed performance metrics, including precision, recall, and F1 score. These reports might be stored in model training logs, documentation, or specialized monitoring tools.
      Verification: Cross-check the calculated F1 score with the one reported in these evaluation documents to ensure consistency.

F. Code and Scripts

Location: Review the scripts or code used for model evaluation to understand the exact methods used for calculating performance metrics.

Verification: Ensure that the methods and formulas used in the scripts align with standard practices.

Appendix 5

Tools to Assess an AI Model's Behavior

TOOL'S OBJECTIVE

Enhancing AI systems' overall transparency, interpretability, explicability, and accountability.

Technique	*Explanation*	*Application*
Alibi	Alibi is an open-source library that provides methods for explaining machine learning models, including anchors, counterfactuals, and SHAP values. These methods help in understanding model decisions and the importance of input features.	Applied to generate diverse explanations for black-box models, Alibi offers real-time insights into model decisions, supporting interpretability, explicability, accountability, audibility, and traceability. This is achieved by providing different explanation techniques suitable for various AI models and scenarios.
Anchors	Produces high-precision explanations by finding conditions (anchors) that sufficiently explain the model's behavior.	Offers clear, rule-based explanations that are easy to understand, supporting interpretability and accountability.
Counterfactual Explanations	Provides explanations by showing how changing input features can alter the prediction outcome.	Useful for illustrating decision boundaries and feature importance, aiding in interpretability and accountability.

(Continued)

Technique	*Explanation*	*Application*
DeepLIFT (Deep Learning Important FeaTures)	Tracks the contribution of each input neuron to the output, providing scores indicating importance.	Useful for interpreting deep learning models, offering insights into feature relevance and supporting traceability and explicability.
Explainable Boosting Machine (EBM)	Uses generalized additive models to provide inherently interpretable and accurate predictions.	Offers transparent and understandable model outputs, supporting interpretability and explicability.
ELI5	ELI5 (Explain Like I'm 5) is a Python library designed to help debug machine-learning classifiers and explain their predictions using techniques such as feature importance, permutation importance, and text explanations.	It supports real-time explanation generation for models such as scikit-learn, XGBoost, and lightGBM. ELI5 enhances interpretability by making complex model predictions understandable, explicability by detailing the decision-making process, accountability by ensuring transparency, audibility by allowing for consistent and clear explanations, and traceability by documenting the explanations and model behavior.
Grad-CAM (Gradient-weighted Class Activation Mapping)	Visualizes the regions in the input that are important for the prediction by using gradients.	Applied primarily in image-based neural networks to highlight important areas, aiding in interpretability and explicability.
Integrated Gradients	Computes the gradients of the prediction with respect to input features, averaging over the path from baseline to input.	Helps in understanding the contribution of each feature to the prediction in neural networks, enhancing interpretability and traceability.
LIME (Local Interpretable Model-agnostic Explanations)	Generates local explanations by approximating the AI model with an interpretable model for individual predictions.	Applied to any black-box model to provide real-time, interpretable explanations for each prediction, enhancing interpretability and accountability.

(Continued)

Technique	*Explanation*	*Application*
Provenance Tracking	Tracks the data and model development lifecycle to ensure traceability and accountability.	Ensures that the origins and transformations of data and models are transparent, supporting traceability and accountability.
RuleFit	Combines decision rules with linear models to create interpretable models that are also predictive.	Provides rule-based explanations for predictions, enhancing accountability and traceability.
SHAP (SHapley Additive exPlanations)	Uses Shapley values from cooperative game theory to attribute the contribution of each feature to the prediction.	Provides consistent and locally accurate explanations for model predictions, improving explicability and traceability.
The What If Tool	This tool provides an interactive interface for exploring AI model performance, analyzing the effects of different features, and diagnosing issues related to fairness and accuracy. It allows users to visually inspect and manipulate data to see how changes affect model predictions.	Provides real-time, visual, and textual explanations by allowing users to manipulate input data and observe prediction changes. This enhances interpretability by making model decisions clear, explicability by providing detailed feature impact analysis, accountability by ensuring model fairness and performance are transparent, audibility by offering consistent and interactive explanations, and traceability by documenting the results of various input manipulations and their effects on the model.

Appendix 6

AI Model Datasheet:[1] Model Name: GPT-4

BASIC INFORMATION

- **Version:** 4.0
- **Release Date:** March 2023
- **Developer:** OpenAI
- **Primary Function:** Natural Language Understanding and Generation
- **Architecture:** Transformer-based language model
- **Supported Languages:** English (primary), additional support for multiple languages

DATASHEET SECTIONS

1. **Intended Use**
    - **Primary Application:** Chatbots, content generation, summarization, code assistance, translation.
    - **Target Users:** Researchers, developers, businesses, educators, and general users.
    - **Limitations:** Not suited for real-time, mission-critical decisions. May generate incorrect or biased information.
2. **Data Information**
    - **Training Data Sources:**
        - **Data Types:** Textual data from books, websites, open data repositories.
        - **Exclusions:** Private or proprietary data, personal identifiers, sensitive content.
        - **Size:** Approximately 570 GB of preprocessed text data.
    - **Preprocessing Details:** Tokenization, deduplication, filtering for harmful content.

3. Model Architecture
   - **Parameters:** 175 billion
   - **Layers:** 96 Transformer layers
   - **Token Limit:** 8,192 tokens per request (input + output combined).
   - **Attention Mechanism:** Scaled dot-product attention with multi-head configuration.
4. Performance Metrics
   - **Evaluation Benchmarks:**
     - GLUE (General Language Understanding Evaluation): High Accuracy
     - SuperGLUE: Top performance across multiple tasks
     - Code Generation: HumanEval Benchmark
   - **Ethical Tests:** Tested for bias, fairness, and harmful content generation.
5. Safety and Ethics
   - **Bias Mitigation:** Trained with efforts to minimize bias; regular updates are applied.
   - Adversarial **Testing:** Resilient against prompt injection and manipulation (within limits).
   - **Limitations:**
     - May still exhibit subtle biases.
     - Requires human oversight for high-stakes decisions.
6. Environmental Impact
   - **Training Compute:** Estimated at several hundred PF-days (Petaflop Days).
   - **Inference Efficiency:** Optimized to reduce latency and compute costs per query.
   - **Carbon Offset Measures:** Commitment to sustainable energy for compute resources.
7. Deployment and Updates
   - **API Availability:** OpenAI API with tiered access.
   - **Versioning:** Regular updates to improve performance and address discovered issues.
   - **User Controls:** Fine-tuning and content moderation options available.
8. Known Issues
   - **Risk of Misuse:** Potential to generate misinformation if not carefully monitored.
   - **Accuracy: Not** guaranteed for real-time, precise knowledge of current events.
   - **Data Dependency:** Training data determines the model's scope and bias.

9. **Contact and Feedback**[2]
   - **Support:** OpenAI Support
   - **Report Issues:** Feedback Form

ENDNOTES

1 The Model Datasheet example provided is synthesized based on publicly available information about GPT-4, combined with general best practices for creating model documentation. The information draws on OpenAI's documentation and guidelines for transparency in AI systems.

2 OpenAI. (2024). GPT-4 Model Datasheet: A synthesized response based on publicly available information and AI documentation standards. Created by ChatGPT, November 2024.

Appendix 7

AI Model Nutrition Label[1]: Model Name GPT-4

MODEL OVERVIEW

- **Name:** GPT-4
- **Developer:** OpenAI
- **Release Date:** March 2023

TRAINING INFORMATION

- **Training Data:** Large-scale text data from publicly available sources (e.g., books, websites)
- **Exclusions:** Private data, personal identifiers, and sensitive content
- **Parameters:** 175 billion
- **Token Limit:** 8,192 tokens (input + output combined).

PERFORMANCE METRICS

- **Language Understanding:**
  - GLUE Score: High Accuracy
  - SuperGLUE: State-of-the-art results
- **Code Generation:** HumanEval Benchmark Top Performer
- **Limitations:** May generate biased or factually incorrect responses.

ETHICS & SAFETY

- **Bias Testing:** Conducted to minimize harmful or biased outputs.
- **Ethical Concerns:**
  - Risk of misinformation.
  - Potential misuse for malicious purposes.

- Safety Measures:
  - Content moderation is available.
  - Human review is recommended for critical use cases.

ENVIRONMENTAL IMPACT

- **Compute for Training:** Several hundred PF-days (Petaflop Days).
- **Efficiency Improvements:** Optimized for reduced inference costs.
- **Sustainability:** OpenAI's commitment to carbon offset compute resources.

DEPLOYMENT

- **Access:** OpenAI API with tiered access levels.
- **Updates:** Regular version improvements based on user feedback.
- **User Controls:** Custom fine-tuning and filtering options are available.[2]

NOTES

1 The Model Nutrition Label example provided is synthesized based on publicly available information about GPT-4, combined with general best practices for creating model documentation. The information draws on OpenAI's documentation and guidelines for transparency in AI systems.

2 OpenAI. (2024). GPT-4 Model Nutrition Label: A synthesized response based on publicly available information and AI transparency standards. Created by ChatGPT, November 2024.

Appendix 8

AI Model Data Statement[1]: Model Name GPT-4

DATA SOURCES

- Origin:
  - GPT-4 was trained on a diverse range of publicly available text data, including:
    - Books
    - Academic papers
    - Online forums
    - Websites
- Exclusions:
  - Proprietary or private datasets.
  - Sensitive or personal identifiers.
  - Any content violating ethical or privacy standards.

DATA TYPES

- Textual Content:
  - General language data for linguistic diversity.
  - Domain-specific texts to enhance understanding in specialized areas.
- Data Formats:
  - Cleaned and tokenized plain text.

PREPROCESSING AND FILTERING

- Steps Taken:
  - **Deduplication:** Removal of repeated or redundant entries.

    - **Normalization:** Ensuring consistent data formatting.
    - **Filtering:** Eliminating harmful or explicit content.
- Techniques:
    - Rule-based content moderation.
    - Human review of critical cases.

ETHICAL CONSIDERATIONS

- Bias Mitigation:
    - Efforts were made to reduce systemic biases in training data.
    - Ongoing evaluation and updates to address identified issues.
- Inclusion:
    - Datasets curated to represent diverse linguistic and cultural perspectives.
- Risk Ma3nagement:
    - Exclusions of hate speech, misinformation, and exploitative content.

DATA LIMITATIONS

- Representation Gaps:
    - Underrepresentation of certain languages, cultures, or perspectives due to data availability.
- Potential Bias:
    - GPT-4's output may reflect biases inherent in the training data.
- Temporal Relevance:
    - Data may not inclSude events or knowledge post-2021.

PRIVACY AND SECURITY

- User Data:
    - No personal user data was included in the training.
    - Strict compliance with data privacy regulations.
- Content Usage:
    - Generated outputs are shaped by public training data, avoiding unauthorized use of copyrighted materials.

DATA IMPACT

- Utility:
    - Supports diverse applications, from education to creative content generation.

- **Limitations:**
  - Requires user oversight to ensure responsible usage.[2]

NOTES

1 The Model Data Statement example provided is synthesized based on publicly available information about GPT-4, combined with general best practices for creating model documentation. The information draws on OpenAI's documentation and guidelines for transparency in AI systems.
2 OpenAI. (2024). GPT-4 Model Data Statement: A synthesized response based on publicly available information and AI documentation practices. Created by ChatGPT, November 2024.

Appendix 9

AI Model Factsheet: Model GPT-4[1]

MODEL OVERVIEW

- **Model Name:** GPT-4
- **Developer:** OpenAI
- **Version:** 4.0
- **Release Date:** March 2023
- **Model Type:** Large-scale Transformer-based Language Model
- **Primary Function:**
  - Natural Language Understanding
  - Text Generation
  - Conversational AI

INTENDED USE

- **Applications:**
  - Chatbots
  - Educational Assistance
  - Content Generation (e.g., blog posts, summaries, creative writing)
  - Translation
  - Code Assistance
- **User Groups:** Developers, researchers, educators, businesses, and general users.
- **Exclusions:**
  - Not suitable for real-time or mission-critical applications.
  - Requires human oversight for high-risk tasks.

MODEL TRAINING

- **Training Data:**
  - Publicly available textual datasets (books, websites, open repositories).
  - Exclusions: Sensitive content, personal data, and proprietary materials.
- **Training Techniques:**
  - Large-scale unsupervised learning.
  - Optimization using autoregressive predictive methods.
- **Parameters:** 175 billion
- **Token Limit:** 8,192 tokens per interaction (input + output).

PERFORMANCE

- **Benchmark Testing:**
  - GLUE: High language understanding accuracy.
  - SuperGLUE: State-of-the-art task performance.
  - HumanEval Benchmark: Top results in code generation.
- **Known Strengths:**
  - Produces coherent, context-aware responses.
  - Excels in multi-turn conversations and complex queries.
- **Known Weaknesses:**
  - Prone to generating factually incorrect or biased responses.
  - May struggle with extremely domain-specific or technical content.

SAFETY AND ETHICAL CONSIDERATIONS

- **Bias Mitigation:**
  - Efforts to reduce systemic biases in training data.
  - Ongoing updates and monitoring for ethical compliance.
- **Risks:**
  - Potential for misuse in generating harmful or misleading content.
  - Requires user discretion for sensitive applications.
- **Safety Mechanisms:**
  - Content filtering.
  - User moderation controls.

ENVIRONMENTAL IMPACT

- **Compute Requirements:** Several hundred PF-days (Petaflop Days) for training.
- **Inference Efficiency:** Optimized to reduce energy consumption during usage.

- **Sustainability:** Commitment to offsetting carbon emissions for computing operations.

DEPLOYMENT

- **Access:** OpenAI API with tiered plans for individual and enterprise users.
- **Customization:** Support for fine-tuning and tailored solutions.
- **Updates:** Regular improvements based on user feedback and research.

LIMITATIONS

- **Knowledge Cutoff:** September 2021.
- **Output Variability:** Can produce verbose or off-topic responses if poorly prompted.
- **Context Sensitivity:** Performs best with clear and concise input queries.[2]

NOTES

1 The Model Factsheet example provided is synthesized based on publicly available information about GPT-4, combined with general best practices for creating model documentation. The information draws on OpenAI's documentation and guidelines for transparency in AI systems.

2 OpenAI GPT-4 Model Factsheet. Synthesized response based on publicly available information and model documentation standards. Created by [ChatGPT], November 2024.

Appendix 10

AI Model (GPT-4) System Card[1]: Model Overview

GPT-4 is a state-of-the-art language model developed by OpenAI. It utilizes deep learning techniques to process and generate human-like text across various topics. GPT-4 is trained on diverse datasets and designed to support applications in research, education, entertainment, and more while maintaining robust safeguards to prevent misuse.

KEY CAPABILITIES

- **Natural Language Understanding:** Superior ability to comprehend and respond to human language in diverse contexts, including casual conversation, technical discussions, and creative writing.
- **Text Generation:** Generates coherent, contextually relevant, and engaging text for tasks such as writing, summarization, translation, and code generation.
- **Multimodal Reasoning:** Limited support for combining textual inputs with structured data (e.g., tables, figures).
- **Domain Adaptability:** Effective across many specialized domains (e.g., science, law, business) due to pretraining on diverse datasets.

ETHICAL CONSIDERATIONS

- OpenAI prioritizes the ethical deployment of GPT-4, including efforts to minimize harmful outputs and address misinformation, bias, and misuse concerns.

BIAS MITIGATION

- The model is trained on a mixture of datasets to ensure fairness, but it can still reflect biases present in training data. Continuous efforts are made to identify and reduce such biases.

SAFETY MEASURES

- Implementations of GPT-4 include safeguards to minimize the generation of harmful, violent, or inappropriate content.
- GPT-4 avoids engaging in topics where misinformation could lead to harm, such as medical, financial, or legal advice, without proper disclaimers.

MISINFORMATION PREVENTION

- The model provides accurate responses when possible but may sometimes produce incorrect or outdated information. Users are encouraged to verify critical information.

KNOWN LIMITATIONS

- **Factual Accuracy:** GPT-4 occasionally produces plausible but incorrect or nonsensical answers, especially when dealing with ambiguous or rare queries.
- **Context Handling:** The model has a limited memory window, which may result in loss of context over long interactions.
- **Ambiguity and Misinterpretation:** It may misinterpret nuanced or ambiguous inputs, leading to unexpected outputs.
- **Lack of Common-Sense Reasoning:** While significantly improved, GPT-4 may still struggle with common-sense reasoning in certain complex scenarios.

DEPLOYMENT AND USE CASES

Recommended Use Cases

- Education and research support
- Creative writing and storytelling
- Summarization of documents
- Code explanation and generation
- Language translation and practice

Discouraged or Restricted Uses

- Any application involving the dissemination of false or harmful information
- Uses that promote violence, discrimination, or harassment
- Sensitive domains (e.g., medical or financial advice) without expert oversight

Ongoing Improvements

- OpenAI actively seeks user feedback to enhance GPT-4's performance and safety. Updates are periodically applied to address emerging challenges, incorporate new knowledge, and refine the model's behavior.

User Responsibility

- While GPT-4 is a powerful tool, users bear responsibility for ensuring its appropriate use. Users should:
  - Verify critical outputs independently.
  - Avoid using the model for harmful purposes.
  - Report unsafe or biased outputs to OpenAI.

Acknowledgments

- The development of GPT-4 benefited from extensive research, user feedback, and contributions from the broader AI community. OpenAI thanks all individuals and organizations who supported this effort.[2]

NOTES

1 The example AI model System Card provided is a synthesized example of a system card for GPT-4, inspired by OpenAI's documentation style and principles. It is not directly sourced from an official OpenAI publication but is informed by general knowledge of OpenAI's practices and transparency reports.

Access to the official system card or detailed documentation, the reader may refer to OpenAI's official publications and resources, such as the OpenAI Model Card for GPT-4 https://openai.com/research/gpt-4.

2 OpenAI, March 14, 2023, "GPT-4 System Card," https://openai.com/index/gpt-4-research/ and GPT-4 Technical Report, March 4, 2024, https://arxiv.org/pdf/2303.08774.

Appendix 11

AI, Audit, and Risk Management: Standards, Frameworks, Guidelines, Toolkits

GENERAL TECHNOLOGY STANDARDS BODIES

ISO/IEC (**International Organization for Standardization / International Electrotechnical Commission**):

- ISO/IEC 22989: Artificial Intelligence – Concepts and Terminology
- ISO/IEC 23053: Framework for Artificial Intelligence (AI) Systems Using Machine Learning
- ISO/IEC TR 24027: Artificial Intelligence – Bias in AI Systems and AI-Aided Decision Making
- ISO/IEC 38507: Governance of IT – Guidance for Artificial Intelligence Governance
- ISO/IEC 24028: Overview of Trustworthiness in Artificial Intelligence

IEEE (**Institute of Electrical and Electronics Engineers**):

- **IEEE 7000 Series:** Ethics and AI, including:
  - IEEE 7001 – Transparency in Autonomous Systems
  - IEEE 7002 – Data Privacy in AI Systems
  - IEEE 7003 – Algorithmic Bias Considerations
- **P2801:** Recommended Practice for AI Transparency

AUDIT AND RISK MANAGEMENT ORGANIZATIONS

ISACA (**Information Systems Audit and Control Association**):

- COBIT 2019: Framework for Governance and Management of Enterprise IT (includes AI governance considerations)
- AI Risk Framework: ISACA guidance on identifying, managing, and auditing risks in AI systems

IIA (**Institute of Internal Auditors**):

- IIA Standard 1210: Proficiency
- IIA Standard 2010: Planning
- IIA Standard 2030: Resource Management
- IIA Standard 2100: Nature of Work
- IIA Standard 2110: Governance
- IIA Standard 2130: Control
- IIA Standard 2200: Engagement Planning
- IIA Standard 2201: Planning Considerations
- IIA Standard 2210: Engagement Objectives
- IIA Standard 2220: Engagement Scope
- IIA Standard 2230: Engagement Resource Allocation
- IIA Standard 2240: Engagement Work Program
- IIA Standard 2310: Identifying Information

COSO (**Committee of Sponsoring Organizations of the Treadway Commission**):

- **ERM Framework:** Enterprise Risk Management (adaptable for AI governance and risk considerations)

AI-SPECIFIC STANDARDS AND FRAMEWORKS

NIST (**National Institute of Standards and Technology**):

- **NIST AI Risk Management Framework (AI RMF 1.0):** Comprehensive framework for assessing and mitigating risks in AI systems
- **NIST AI 600-1:** Artificial Intelligence Risk Management Framework: Generative Artificial Intelligence Profile
- **NIST SP 1270:** Towards a Standard for Identifying and Managing Bias in AI
- **NIST SP 500-290:** Big Data Interoperability Framework, including AI considerations

Partnership on AI:

- **AI Fairness, Transparency, and Accountability Guidelines:** Best practices for implementing AI ethics in organizations

OECD (**Organisation for Economic Co-operation and Development**):

- **OECD AI Principles:** High-level principles addressing AI accountability, transparency, and fairness

PROFESSIONAL BODIES IN ETHICS AND GOVERNANCE

ACM (Association for Computing Machinery):

- **ACM Code of Ethics and Professional Conduct:** Ethical considerations for AI professionals
- **AI Ethics Guidelines:** Focused on accountability and transparency in AI systems

IFAC (International Federation of Accountants):

- **International Standard on Assurance Engagements (ISAE) 3000:** Guidelines for AI-related assurance and audit services

ISO 37001 Alliance (Ethics and Compliance Standards):

- **ISO 37001 (Anti-Bribery):** Includes considerations for AI-driven compliance tools

CROSS-SECTORAL AI AND DATA-FOCUSED ORGANIZATIONS

World Economic Forum (WEF):

- **AI Procurement Guidelines:** Framework for ethical and risk-based procurement of AI solutions

Global Partnership on AI (GPAI):

- **AI for the Public Good Standards:** Guidelines for balancing AI innovation and social responsibility

ENISA (European Union Agency for Cybersecurity):

- **AI Threat Landscape Report:** Comprehensive overview of AI risks and threat mitigation strategies

ISO 31000 (Risk Management Framework):

- **Adaptation for AI-Specific Risks:** Covers guidelines for managing emerging AI risks

MITRE Adversarial Threat Landscape for Artificial-Intelligence Systems (ATLAS): a globally accessible, living knowledge base of adversary tactics and techniques based on real-world attack observations and realistic demonstrations from artificial intelligence (AI) red teams and security groups (https://atlas.mitre.org/).

IBM AI Adversarial Robustness 360: The open-source Adversarial Robustness Toolbox provides tools that enable developers and researchers to evaluate and defend machine learning models and applications against the adversarial threats of evasion, poisoning, extraction, and inference (https://research.ibm.com/projects/adversarial-robustness-toolbox).

IBM AI Fairness 360: This extensible open-source toolkit can help you examine, report, and mitigate discrimination and bias in machine learning models throughout the AI application lifecycle (https://research.ibm.com/blog/ai-fairness-360).

IBM AI Privacy 360: This toolbox includes several tools to support the assessment of privacy risks of AI-based solutions and to help them adhere to any relevant privacy requirements. Tradeoffs between privacy, accuracy, and performance can be explored at different stages in the ML lifecycle (https://research.ibm.com/blog/ai-privacy-toolkit).

IBM AI Explainability 360
This extensible open-source toolkit can help you comprehend how machine learning models predict labels by various means throughout the AI application lifecycle (https://research.ibm.com/blog/ai-explainability-360).

SECTOR-SPECIFIC STANDARDS

Financial Industry:

- **Basel Committee on Banking Supervision (BCBS):** Principles for AI use in banking risk management
- **Financial Industry Regulatory Authority (FINRA):** AI-specific risk alerts and guidelines for algorithmic trading

Healthcare Industry:

- **Health Level Seven International (HL7):** AI standards for healthcare data interoperability
- **ISO 13485:** Adapted for AI-based medical device governance

Transportation:

- **SAE International:** Standards for AI in autonomous vehicles, including SAE J3016

SPECIALIZED AI ETHICS AND BIAS ENTITIES

AI Now Institute:

- **AI Accountability Framework:** Specific guidance for AI audits related to bias, accountability, and transparency

Ethics Guidelines for Trustworthy AI (EU High-Level Expert Group on AI):

- **Trustworthy AI Principles:** Framework emphasizing accountability, fairness, and transparency

Appendix 12

AI Application Datasheet Example[1]

SYSTEM OVERVIEW

Name of AI System: Sentiment Analyzer Pro 2.0

Purpose: To analyze and categorize the sentiment (positive, neutral, negative) of text data in real-time, designed for use in social media monitoring, customer feedback, and market research.

Primary Use Case: Social media sentiment analysis for brand management.

Developer: AI Insights Inc.

Version: 2.0

Release Date: January 2024

TRAINING AND DATASET INFORMATION

Training Data Source: Open-text datasets including Twitter, Reddit, and public product reviews.

Dataset Size: 50 million text samples.

Preprocessing: Text normalization, removal of stop words, and tokenization.

Bias Mitigation: Balanced dataset across sentiments and demographics to minimize representational bias.

Known Limitations: Limited performance on slang or domain-specific jargon.

PERFORMANCE METRICS

Overall Accuracy: 91%
Precision: 90%
Recall: 88%
F1 Score: 89%
Benchmark Scenarios: Tested on datasets with varying text lengths and diverse topics.
Performance may degrade with highly ambiguous sentences.

ETHICAL CONSIDERATIONS

Privacy Measures: All training data is anonymized and sourced from public datasets.
Fairness: Tested across multiple demographic groups to ensure balanced sentiment predictions.
Risk Mitigation: Regular audits to identify and mitigate potential biases in new versions.

COMPLIANCE AND GOVERNANCE

Regulatory Compliance: Adheres to GDPR data privacy standards and ISO 27001 for information security.
Third-Party Audits: Independently audited by Ethical AI Labs in December 2023.
Certification: Certified under the NIST AI Risk Management Framework.

USER AND DEPLOYMENT INFORMATION

Installation: Available as a cloud-based API and on-premise software.
System Requirements: Minimum 8GB RAM, 4-core CPU, Python 3.8+.
Usage Instructions: Input text data through the API; outputs sentiment category and confidence score.
Support Contact: support@aiinsights.com.

MAINTENANCE AND UPDATES

Update Policy: Updates provided quarterly, including retraining with newer datasets.

Version Control: Version history and changelogs are available on the developer's website.

Issue Reporting: Users can report issues via the support portal at https://www.aiinsight.com/contact-us/.

NOTE

1 Assistant, December 11, 2024. "AI Application Data Sheet Example: Sentiment Analyzer Pro 2.0.," Sentiment Analyzer Pro 2.0 is a fictitious product. The Data Sheet document was designed and produced for educational and illustrative purposes via author prompt with OpenAI v4o.

References

These additional materials were read, reviewed, and consulted in preparing this book.

(n.a.), (n.d.), "AI Ethics Principles & Guidelines, The Smart Dubai Office, www.smart-dubai.ae/docs/default-source/aiprinciples-resources/ai-ethics.pdf?sfvrsn=d4184f8d_6.

(n.a.) (n.d.), "Change Management, Enterprise Performance Life Cycle Framework, Practices Guide v1.0," Department Of Health And Human Services, www.hhs.gov/sites/default/files/ocio/eplc/EPLC%20Archive%20Documents/07%20-%20Change%20Management%20Plan/eplc_change_management_practices_guide.pdf.

(n.a.), (n.d.), "The Four Essential AI Use Cases," Yext, statista.tradepub.com/free/w_yext08/prgm.cgi.

(n.a.), (n.d.), "The Critical Importance of the Reward Function in Reinforcement Learning," www.alphanome.ai/post/the-critical-importance-of-the-reward-function-in-reinforcement-learning#:~:text=The%20reward%20function%20is%20a,and%20behavior%20of%20the%20agent.

(n.a.), (n.d.), "The Future of AI in Technology Trends and Transformations," https://statista.tradepub.com/?p=w_accx55&w=d&email=c12b5331a06a8f4ee04668a880bdeb14&key=n7mHGh2jT5r2WNPyzpqP&ts=13433&u=07420709404916940131328&e=bWFyY2VsYUB3ZWJzdGVyLmVkdQ==&secure=1&_afn=0.

(n.a.), September 2024, *"Change Management Policy v1.2,"* Wichita State University, www.wichita.edu/services/its/documents/ChangeManagementRevised.pdf.

(n.a.), October 30, 2023, "Hiroshima Process International Code of Conduct for Organizations Developing Advanced AI Systems," www.mofa.go.jp/files/100573473.pdf.

(n.a.), October 27, 2023, *"Emerging Processes for Frontier AI Safety,"* Government of the UK, Department for Science, Innovation and Technology, AI Safety Summit 2023, www.gov.uk/government/publications/emerging-processes-for-frontier-ai-safety

(n.a.), August 2, 2023, "With AI, you get what you give," Thomson Reuters, Casetext, https://casetext.com/blog/prompt-engineering-best-ai-output/

(n.a.), May 12, 2023, "The 5 Steps of Reinforcement Learning with Human Feedback,", www.appen.com/blog/the-5-steps-of-reinforcement-learning-with-human-feedback#:~:text=Reward%20model%20training%20is%20an,accuracy%20to%20the%20desired%20outcome.

(n.a.), April 27, 2023, "Auditing machine learning algorithms," The Supreme Audit Institutions of Finland, Germany, the Netherlands, Norway and the UK, www.auditingalgorithms.net.

(n.a.), November 25, 2022, "Change Management," Navvia, https://navvia.com/hubfs/Davids%20Files/Navvia%20-%20Change%20Management%20Process%20-%2025Nov2022.pdf

(n.a.), September 23, 2022, "Auditing algorithms: the existing landscape, role of regulators and future outlook," Digital Regulation Corporation Forum, www.gov.uk/government/publications/findings-from-the-drcf-algorithmic-processing-workstream-spring-2022/auditing-algorithms-the-existing-landscape-role-of-regulators-and-future-outlook.

(n.a.), February 15, 2022, *"Change Management Process,"* University of Cincinnati, www.uc.edu/content/dam/uc/ucit/docs/so/cm_process_document.pdf.

Accenture, June 2022, "From AI Compliance to Competitive Advantage: Becoming Responsible by Design", www.accenture.com/content/dam/accenture/final/a-com-migration/r3-3/pdf/pdf-179/accenture-responsible-by-design-report.pdf.

Adkins, D., et al., October 17, 2022, "Method Cards for Prescriptive Machine-Learning Transparency," In *IEEE/ACM 1st International Conference on AI Engineering – Software Engineering for AI (CAIN)*, IEEE, Pittsburgh, PA, 2022, 90–100 pages. doi: 10.1145/3522664.3528600.

AI Incident Database. https://incidentdatabase.ai.

Alber, S., February 9, 2022, *"Directed Acyclic Graphs (DAGs) and Regression for Causal Inference,"* UC Davis Health, https://health.ucdavis.edu/media-resources/ctsc/documents/pdfs/directed-acyclic-graphs20220209.pdf.

Arnold, M., et al., July-September 2019, "FactSheets: Increasing Trust in AI Services Through Supplier's Declarations of Conformity," In *IBM Journal of Research and Development*, vol. 63, no. 4/5, pp. 6:1–6:13, https://krvarshney.github.io/pubs/ArnoldBHHMMNROPRRTV_ijrd2019.pdf.

Bader, M., Beddor, J., Bothwell, B., and Businsky, B., et al., September 28, 2020, "Agile Assessment Guide: Best Practices for Agile Adoption and Implementation," *U.S. Government Accountability Office*, GAO-20-590G, www.gao.gov/products/gao-20-590g.

Badyal, N. et al., 2023, "Intentional Biases in LLM Responses," arXiv. https://arxiv.org/pdf/2311.07611

Barney, N., and Schmelzer, R., July 28, 2023, "6 Ways to Reduce Different Types of Bias in Machine Learning," www.techtarget.com/searchenterpriseai/feature/6-ways-to-reduce-different-types-of-bias-in-machine-learning?utm_campaign=20230818_Machine+learning+explained&utm_medium=email&utm_source=MDN&asrc=EM_MDN_274892125&bt_ee=gl8dDNg9NVjPNsO7JNvgMGMmmlY0x5Wo%2BgGJQj8NP8cTsQgQ3bM0vovAvRctAd36&bt_ts=1692626742462.

Bengio, Y., et al., October 26, 2023, "Managing AI Risks in an Era of Rapid Progress," AITimes.org, vol. 384, no. 6698, pp. 842–845, https://arxiv.org/abs/2310.17688.

Bowman, S.R., 2023, "Eight Things to Know about Large Language Models," arXiv preprint arXiv:2304.00612.

Burgess, M., 2024, "Generative AI's Biggest Security Flaw Is Not Easy to Fix," WIRED, www.wired.com/story/generative-ai-prompt-injection-hacking.

Canadian Centre for Cyber Security (2023) Generative artificial intelligence (AI) - ITSAP.00.041. www.cyber.gc.ca/en/guidance/generative-artificial-intelligence-ai-itsap00041

Chmielinski, K., Newman, S., Kranzinger, C., and Hind , M., et al., May 2024, "*The CLeAR Documentation Framework for AI Transparency: Recommendations for Practitioners and Context for Policymakers*," The Shorenstein Center on Media, Politics and Public Policy, Harvard Kennedy School, Harvard University, https://shorensteincenter.org/wp-content/uploads/2024/05/CleAR_KChmielinski_FINAL.pdf.

Crockett, K., Colyer, E., Gerber, L., and Latham, A., August 2023, "Building Trustworthy AI Solutions: A Case for Practical Solutions for Small Businesses" in *IEEE Transactions on Artificial Intelligence*, vol. 4, no. 04, pp. 778–791, doi: 10.1109/TAI.2021.3137091.

https://e-space.mmu.ac.uk/629069/1/TAI3137091%20corrected%20final%20version%20colour%20photos.pdf.

Culafi, A., June 7, 2023, "What Generative AI's Rise Means for the Cybersecurity Industry," TechTarget, www.techtarget.com/searchsecurity/news/366539613/What-generative-AIs-rise-means-for-the-cybersecurity-industry.

De Angelo, D., February 1, 2024, "Short, Mid and Long-Term Impacts of AI in Cybersecurity," Palo Alto Networks, www.paloaltonetworks.com/blog/2024/02/impacts-of-ai-in-cybersecurity.

Department of Technology and Innovation, October 6, 2022, "Change Management Standard," City of Albuquerque, https://eweb.cabq.gov/tools/IT%20Policies%20Standards%20and%20Procedures/Change%20Management%20Standard%20-%20Final.pdf.

El-Kishky, A., Selsam, D., and Song, F., et al., September 12, 2024, "OpenAI o1 System Card," *OpenAI*, https://cdn.openai.com/o1-system-card-20240917.pdf.

Fang, H., and Miao, H., July 29, 2020, "Introducing the Model Card Toolkit for Easier Model Transparency Reporting," Google Research, https://blog.research.google/2020/07/introducing-model-card-toolkit-for.html.

Ford, J., October 18, 2024, "*Showing AI Users Diversity in Training Data Boosts Perceived Fairness and Trust,*" Penn State University, www.psu.edu/news/research/story/showing-ai-users-diversity-training-data-boosts-perceived-fairness-and-trust.

General Data Protection Regulation, https://eur-lex.europa.eu/eli/reg/2016/679/oj.

González, F., Ortiz, T., and Ávalos, R., 2020, "Responsible Use of AI for Public Policy: Data Science Toolkit," A Joint Report by the International Development Bank (IDB) and the OECD, https://publications.iadb.org/publications/english/document/Responsible-use-of-AI-for-public-policy-Data-science-toolkit.pdf.

Green, N., Cheema, A., Procope, C., and Adediji, A., February 23, 2022, "System Cards, a New Resource for Understanding How AI Systems Work," https://ai.meta.com/blog/system-cards-a-new-resource-for-understanding-how-ai-systems-work.

Gursoy, F., and Kakadiaris, I., June 28, 2022, "*System Cards for AI-Based Decision-Making for Public Policy,*" Montreal Ai Ethics Institute, https://montrealethics.ai/system-cards-for-ai-based-decision-making-for-public-policy.

Hammer, A., February 8, 2024, "Understanding Fair AI: A Simple Guide!", Readynez, www.readynez.com/en/blog/understanding-fair-ai-a-simple-guide/#:~:text=To%20make%20AI%20models%20fair,crucial%20for%20improving%20AI%20fairness.

Heger, A., Marquis, L., and Vorvoreanu, M., et al., November 2022, "Understanding Machine Learning Practitioners' Data Documentation Perceptions, Needs, Challenges, and Desiderata," *Microsoft, 2022 Conference on Computer Supported Cooperative Work*, www.microsoft.com/en-us/research/publication/understanding-machine-learning-practitioners-data-documentation-perceptions-needs-challenges-and-desiderata.

Hendrycks, D., Mazeika, M., and Woodside, T., October 9, 2023, "An Overview of Catastrophic AI Risks," https://arxiv.org/pdf/2306.12001.pdf.

High-Level Expert Group on AI, April 8, 2019, "ETHICS GUIDELINES FOR TRUSTWORTHY AI, High-Level Expert Group on Artificial Intelligence," European Commission, https://digital-strategy.ec.europa.eu/en/library/ethics-guidelines-trustworthy-ai.

Hind, M., Mehta, S., Mojsilovíc, A., Nair, R., Ramamurthy, K.N., and Olteanu, A., et al, February 7, 2019, "Increasing Trust in AI Services through Supplier's Declarations of Conformity," https://arxiv.org/pdf/1808.07261.

Information Technology Industry Council, January 2024, "Authenticating AI-Generated Content," www.itic.org/policy/ITI_AIContentAuthorizationPolicy_122123.pdf.

Johnson, M.K., Hanna, Michael, M., Clemens-Sewall, M.V., and Staheli, D.P., 2023. *Responsible AI Toolkit (RAI Toolkit 1.0), Responsible AI, US Department of Defense*, Arlington, VA, [online], https://rai.tradewindai.com.

Karras, T., Laine, S., and Aila, T. 2020, "Analyzing and Improving the Image Quality of StyleGAN," In *Proceedings of the IEEE/CVF Conference on Computer Vision and Pattern Recognition (CVPR)*, 2020, 8110–8119 pages. https://arxiv.org/abs/1912.04958

KPMG, January 2024, "KPMG Generative AI Consumer Trust Survey," https://kpmg.com/kpmg-us/content/dam/kpmg/corporate-communications/pdf/2024/kpmg-generative-ai-consumer-trust-survey.pdf.

Koch, R., 2024, "The Value of Data Quality and Data Diversity in AI models," Clickworker, www.clickworker.com/customer-blog/data-quality-and-diversity/#:~:text=Meaning%20of%20data%20diversity%20in%20AI%20training%20datasets,-When%20training%20AI&text=Allows%20the%20AI%20model%20to,the%20accuracy%20of%20the%20model.

Kusnezov, D., et al., June 15, 2023, "Risks and Mitigation Strategies for Adversarial Artificial Intelligence Threats: A DHS S&T Study," Preparedness Series, Department of Homeland Security, www.dhs.gov/sites/default/files/2023-12/23_1222_st_risks_mitigation_strategies.pdf.

Lawrence, C., Cui, I., and Ho, D., August 8, 2023, "The Bureaucratic Challenge to AI Governance: An Empirical Assessment of Implementation at U.S. Federal Agencies.," In *AAAI/ACM Conference on AI, Ethics, and Society (AIES '23)*, August 08–*10, 2023*, Montréal, QC, Canada. ACM, New York, 47 pages https://doi.org/10.1145/3600211.3604701

Lawton, G., August 22, 2023, "8 Areas for Creating and Refining Generative AI Metrics," TechTarget, www.techtarget.com/searchenterpriseai/feature/Areas-for-creating-and-refining-generative-AI-metrics?utm_campaign=20230824_

ERU-ACTIVE_WITHIN_90_DAYS&utm_medium=email&utm_source=SGERU&source_ad_id=366549253&src=15003291&asrc=EM_SGERU_275384501.

Lawton, G., August 9, 2023, "The Different Types of Machine Learning Explained," TechTarget, www.techtarget.com/searchenterpriseai/tip/Types-of-learning-in-machine-learning-explained?utm_campaign=20230818_Machine+learning+explained&utm_medium=email&utm_source=MDN&asrc=EM_MDN_274892123&bt_ee=gl8dDNg9NVjPNsO7JNvgMGMmmlY0x5Wo%2BgGJQj8NP8cTsQgQ3bM0vovAvRctAd36&bt_ts=1692626742462.

Lee, H. et al., February 10, 2024, "Deepfakes, Phrenology, Surveillance, and More! A Taxonomy of AI Privacy Risks," arXiv. https://arxiv.org/pdf/2310.07879.

Lenaerts-Bergmans, B., March 20, 2024, "Data Poisoning: The Exploitation of Generative AI," Crowdstrike. HI, Honolulu, USA. www.crowdstrike.com/cybersecurity-101/cyberattacks/data-poisoning.

Madaio, M., Stark, L., Vaughan, J.W., and Wallach, H., 2020, "CoDesigning Checklists to Understand Organizational Challenges and Opportunities around Fairness in AI," In *Proceedings of the 2020 CHI Conference on Human Factors in Computing Systems* (CHI 2020), https://doi.org/10.1145/3313831.3376445.

Marinos, T., and Gela, M., 2024, "Responsible AI: The Role of Data and Model Cards," Datatonic, https://datatonic.com/insights/responsible-ai-data-model-cards.

McMillan-Major, A., and Bender, E., 2024, "A Guide for Creating and Documenting Language Datasets with Data Statements - Schema Version 3," Value Sensitive Design Lab, Tech Policy Lab, University of Washington., Seattle, https://techpolicylab.uw.edu/wp-content/uploads/2024/11/Data_Statements-Guide-III.pdf.

Microsoft Corporation, August 25, 2022, "Aether Data Documentation Template," www.microsoft.com/en-us/research/uploads/prod/2022/07/aether-datadoc-082522.pdf.

Microsoft Corporation, June 2022, "Microsoft Responsible AI Impact Assessment Template. Microsoft; https://blogs.microsoft.com/wp-content/uploads/prod/sites/5/2022/06/Microsoft-RAI-Impact-Assessment-Template.pdf.

Moir, S., April 6, 2023, "Documenting Machine Learning Models," This is Important, https://thisisimportant.net/posts/documenting-machine-learning-models.

Mökander, J., and Floridi, L., 2023, "Operationalising AI Governance Through Ethics-Based Auditing: An Industry Case Study," *AI and Ethics* vol. 3, pp. 451–468, https://doi.org/10.1007/s43681-022-00171-7, www.ncbi.nlm.nih.gov/pmc/articles/PMC9152664/pdf/43681_2022_Article_171.pdf.

Moody's, February 2024, "How Can Artificial Intelligence Transform Risk and Compliance?" www.moodys.com/web/en/us/about/insights/data-stories/kyc-ai-risk-and-compliance-survey.html.

Moody's, February 2024, "Navigating the AI Landscape," KYC Solutions at Moody's Analytics, www.moodys.com/web/en/us/site-assets/ma-kyc-navigating-the-ai-landscape-report.pdf

Murphy, J., October 10, 2023, "10 Realistic Business Use Cases for ChatGPT," TechTarget, www.techtarget.com/searchenterpriseai/tip/Realistic-business-use-cases-for-ChatGPT?utm_campaign=20231017_Realistic+business+use+cases+for+ChatGPT&utm_medium=email&utm_source=MDN&asrc=EM_MDN_279694770&bt_ee=VTFHYAblllEi3JdKL3K2CLqh98DtoEwpZzW1xvcX%2BLrf802kdVsFx05ZRIwREQoB&bt_ts=1697637792819.

National Institute of Standards and Technology, January 2023, "AI Risk Management Framework," www.nist.gov/itl/ai-risk-management-framework, https://doi.org/10.6028/NIST.AI.100-1.

Newman, J., January 2023, "A Taxonomy of Trustworthiness for Artificial Intelligence," UC Berkeley Center for Long-Term Cybersecurity (CLTC), https://cltc.berkeley.edu/wp-content/uploads/2023/01/Taxonomy_of_AI_Trustworthiness.pdf.

OECD, May 2024, "Defining AI Incidents and Related Terms" *OECD Artificial Intelligence Papers*, no. 16, OECD Publishing, Paris, www.oecd-ilibrary.org/deliver/d1a8d965-en.pdf?itemId=%2Fcontent%2Fpaper%2Fd1a8d965-en&mimeType=pdf.

OECD. AI 2023, Database of National AI Policies & Strategies, https://oecd.ai/en/dashboards/overview.

OECD 2019, OECD Recommendation of the Council on Artificial Intelligence, OECD/LEGAL/0449, https://legalinstruments.oecd.org/en/instruments/OECD-LEGAL-0449.

OECD, 2019, "Principles on AI," [Online]. Available: https://www.oecd.org/going-digital/ai/principles/

Office of the White House, October 31, 2023, "White House Executive Order on AI," Center for AI Safety, AI Safety Newsletter #25, www.safe.ai.

Open AI, March 10, 2023, "GPT-4 System Card," https://cdn.openai.com/papers/gpt-4-system-card.pdf.

Open AI, September 12, 2024, "OpenAI o1 System Card," https://openai.com/index/openai-o1-system-card.

Ozoani, E., Gerchick, M., and Mitchell, M., 2022, Model Card Guidebook. Hugging Face, https://huggingface.co/docs/hub/en/model-card-guidebook.

Park, P., Goldstein, S., and O'Gara, A., et al., August 28, 2023, "AI Deception: A Survey of Examples, Risks, and Potential Solutions," arXiv preprint arXiv:2308.14752.

Pownall, C., July 2019, "Understanding the Reputational Risks of AI," ResearchGate. www.researchgate.net/publication/340088726_Understanding_the_Reputational_Risks_of_A.

Pratt, M., June 9, 2020, "10 Common Uses for Machine Learning Applications in Business – FRAUD Detection," www.techtarget.com/searchenterpriseai/feature/10-common-uses-for-machine-learning-applications-in-business?utm_campaign=20230818_Machine+learning+explained&utm_medium=email&utm_source=MDN&asrc=EM_MDN_274892124&bt_ee=gl8dDNg9NVjPNsO7JNvgMGMmmlY0x5Wo%2BgGJQj8NP8cTsQgQ3bM0vovAvRctAd36&bt_ts=1692626742462.

Pupillo, L., 22 January 2020, CEPS Task Force on Artificial Intelligence and Cybersecurity Technology, Governance and Policy Challenges Task Force Evaluation of the HLEG Trustworthy AI Assessment List (Pilot Version). CEPS Task Force Report www.academia.edu/91972914/CEPS_Task_Force_on_Artificial_Intelligence_and_Cybersecurity_Technology_Governance_and_Policy_ChallengesTask_Force_Evaluation_of_the_HLEG_Trustworthy_AI_Assessment_List_Pilot_Version_CEPS_Task_Force_Report_22_January_2020?auto=download&auto_download_source=social-news.

Pw, C., April 2024, "US Responsible AI Survey," PricewaterhouseCoopers, www.pwc.com/us/en/tech-effect/ai-analytics/responsible-ai-survey.html.

Qi, X., Zeng, Y., Xie, T., Chen, P. Y., Jia, R., Mittal, P., and Henderson, P., October 5, 2023, "Fine-Tuning Aligned Language Models Compromises Safety, Even When Users Do Not Intend To!" arXiv:2310.03693v1 [cs.CL], https://arxiv.org/abs/2310.03693.

Reagan, M., May 24, 2021, "Understanding Bias and Fairness in AI Systems" *Towards Data Science*, https://towardsdatascience.com/understanding-bias-and-fairness-in-ai-systems-6f7fbfe267f3.

Schuett, J., 2024, "Frontier AI Developers Need an Internal Audit Function" *Risk Analysis*, pp. 1–21. https://doi.org/10.1111/risa.17665.

Schuett, J, May 26, 2023, "AGI Labs Need an Internal Audit Function," Centre for the Governance of AI, https://arxiv.org/pdf/2305.17038.pdf.

Scott, I., Carter, S., and Coiera, E., January 12, 2021, "Clinician Checklist for Assessing Suitability of Machine Learning, Applications in Healthcare," *BMJ Health Care Inform*, vol. 28, e100251. doi:10.1136/bmjhci-2020-100251.

Sébastien, D., Fischer, F., and Frija, R., October 2018, "Digital Ethics A Guide For Professionals Of The Digital Age," www.cigref.fr/wp/wp-content/uploads/2019/02/Cigref-Syntec-Digital-Ethics-Guide-for-Professionals-of-Digital-Age-2018-October-EN.pdf.

Seifert, C., Scherzinger, S., and Wiese, L., 2019, "Towards Generating Consumer Labels for Machine Learning Models," In IEEE First International Conference on Cognitive Machine Intelligence (CogMI), 173–179 pages, https://research.utwente.nl/files/158031484/Seifert2019_cogmi_consumer_labels_preprint.pdf.

Simon, J., March 18, 2024, "Can AI be Sued for Defamation?", *Columbia Journalism Review, Journalism Protection Initiative, Craig Newmark Graduate School of Journalism*, www.cjr.org/anal-ysis/ai-sued-suit-defamation-libel-chatgpt-google-volokh.php.

Tagliabue, J., et al., October 24, 2021, "DAG Card is the New Model Card," 35th Conference on Neural Information Processing Systems (NeurIPS 2021), Sydney, Australia, https://arxiv.org/pdf/2110.13601.

The White House, Presidential Actions, Biden. J., October 30, 2023, "Executive Order on the Safe, Secure, and Trustworthy Development and Use of Artificial Intelligence," https://bidenwhitehouse.archives.gov/briefing-room/presidential-actions/2023/10/30/executive-order-on-the-safe-secure-and-trustworthy-development-and-use-of-artificial-intelligence/.

The White House, Office of Science and Technology Policy (OSTP), "Blueprint for an AI Bill of Rights," https://bidenwhitehouse.archives.gov/ostp/ai-bill-of-rights/what-is-the-blueprint-for-an-ai-bill-of-rights/

TrustArc, February 19, 2024, "Responsible AI Certification Assessment Criteria, v1.0," https://trustarc.com/wp-content/uploads/2024/03/Responsbile-AI-Certification-Assessment-Criteria.pdf.

UK Government 2022, National AI Strategy, https://www.gov.uk/government/publications/national-ai-strategy.

UK Parliament 2023, The Data Protection and Digital Information (No. 2) Bill 2022-2023, https://commonslibrary.parliament.uk/research-briefings/cbp-9746/.

US Congress 2022, Algorithmic Accountability Act (AAA), https://www.congress.gov/bill/117th-congress/house-bill/6580.

Valentine, C., May 3, 2024, "Data Sets & Diversity: What's the Connection?" CoVABiz, https://covabizmag.com/data-sets-diversity-whats-the-connection.

Vassilev, A., Oprea, A., and Fordyce, A., January 2024, “Adversarial Machine Learning: A Taxonomy and Terminology of Attacks and Mitigations”, National Institute of Standards and Technology, https://csrc.nist.gov/pubs/ai/100/2/e2023/final.

Wernert, R., May 10, 2024, “What is a DAG?”, Summit Mining Ltd., www.summit.io/blog-posts/what-is-a-dag.

Yadav, A., September 8, 2024, “Reward Function in Reinforcement Learning Biased-Algorithms,”Biased-Algorithms,https://medium.com/biased-algorithms/reward-function-in-reinforcement-learning-c9ee04cabe7d#:~:text=In%20simple%20terms%2C%20a%20reward,function%20hands%20over%20a%20reward.

Zhang, W., September 22, 2024, “AI Fairness in Practice: Paradigm, Challenges, and Prospects,” *AI Magazine*, vol. 45, no. 3, pp. 386–395, https://onlinelibrary.wiley.com/doi/pdfdirect/10.1002/aaai.12189?download=true.

Glossary

Additional resources referenced in the compilation of this glossary include but are not limited to the following documents.

Accountability: An assurance that an individual or organization is evaluated on its performance or behavior related to something for which it is responsible.

Adversarial: The term "adversarial" comes from the two competing networks creating and discerning content – a generator network and a discriminator network. For example, in an image-generation use case, the generator network creates new images that look like faces. In contrast, the discriminator network tries to tell the difference between authentic and generated images. The discriminator performance data then helps to train the overall system.

Adversarial Machine Learning: Adversarial Machine Learning is like a game between two smart players. Imagine you have a smart computer program (let's call it the "Defender") that is trained to do a task, like recognizing pictures of cats and mice. Now, another sneaky computer program (we'll call it the "Attacker") tries to fool the Defender by showing it tricky images.

The Attacker's goal is to create special pictures that look confusing, maybe a mix of cat and mice features, or with small changes that are hard for humans to notice. When the Defender sees these pictures, it might make mistakes and think a cat is a mouse or vice versa.

The Defender learns from its mistakes and tries to get better at telling cats and mice apart, while the Attacker keeps trying new ways to create tricky pictures. It is like a never-ending game of cat-and-mouse!

Adversarial Machine Learning is about studying these sneaky attacks and making the Defender more robust, so it can handle the tricky pictures better and make fewer mistakes.

AI Agent: A system that perceives its environment, makes decisions, and takes actions to achieve specific goals. It uses algorithms to process inputs (like data or observations) and determines the best actions to maximize success, often based on feedback or rewards. Examples include virtual assistants, self-driving cars, and game-playing programs.

AI Disaster: A serious AI incident that disrupts the functioning of a community, a society, or an organization, and that may test or exceed its capacity to cope using its own resources. The effect of an AI disaster can be immediate and localized or widespread and lasting for a long period of time.[1]

AI Fairness: A foundational principle in artificial intelligence that aims to guarantee just and equitable outcomes for individuals or groups, irrespective of their diverse characteristics. This concept addresses the potential biases and discriminatory impacts that AI models may unintentionally perpetuate, such as those based on gender, race, or age.[2]

AI Hazard: An event, circumstance, or series of events where the development, use, or malfunction of one or more AI systems could plausibly lead to an AI incident, i.e., any of the following harms:

a) Injury or harm to the health of a person or groups of people.
b) Disruption of the management and operation of critical infrastructure.
c) Violations of human rights or a breach of obligations under the applicable law intended to protect fundamental, labor, and intellectual property rights.
d) Harm to property, communities, or the environment.[3]

AI Incident: An event, circumstance, or series of events where the development, use, or malfunction of one or more AI systems directly or indirectly leads to any of the following harms:

a) Injury or harm to the health of a person or groups of people.
b) Disruption of the management and operation of critical infrastructure.
c) Violations of human rights or a breach of obligations under the applicable law intended to protect fundamental, labor, and intellectual property rights.
d) Harm to property, communities, or the environment.[4]

AI Practitioners: Are all individuals or organizations that develop (including research, design, or provide data for) deploy (including implement), or use AI systems, excluding those that use AI systems in the capacity of end-user or consumer?[5]

AI System: A machine-based system designed to operate with varying levels of autonomy and that may exhibit adaptiveness after deployment. For explicit or implicit objectives, it infers, from the input it receives, how to generate outputs such as predictions, content, recommendations, or decisions that can influence physical or virtual environments.[6]

Algorithm: A finite sequence of instructions, which are followed by a computer system, which is designed to complete a task.

Algorithmic Auditing: A process for verifying that decision-making algorithms produce the expected outcomes without violating legal or ethical parameters.

Algorithmic Fairness: Mathematical representation of a specific definition of fairness that is incorporated into the model selection and fitting process. It is important to consider that different definitions can be exclusive; that is, satisfying one may simply not satisfy the others.[7]

Algorithmic Inequality: Technical flaws in the models that produce a disparity of results for protected groups and that must be evaluated under the definition of algorithmic fairness previously determined.

Algorithm-of-Thought Prompting: Algorithm-of-thought prompting involves guiding an AI model to follow specific, structured algorithms or methodologies when processing information or solving problems. This technique aims to improve the model's consistency and accuracy by embedding well-defined logical or procedural frameworks into its reasoning process.

Alignment: This involves ensuring that powerful AI is properly aligned with human values. The challenge of alignment has two parts. The first part is technical and focuses on how to formally encode values or principles in artificial agents so that they reliably do what they ought to do. The second part of the value alignment question is normative. It asks what values or principles, if any, we ought to encode in artificial agents.

Amazon Bedrock (AWS Bedrock): It is a machine learning platform used to build generative artificial intelligence (AI) applications on the Amazon Web Services cloud computing platform. Bedrock uses foundation models to simplify the creation of these apps and make the process more efficient.

Analytics: It is the application of scientific and mathematical methods to the study and analysis of problems involving complex systems.

There are three distinct types of analytics:

- Descriptive Analytics gives insight into past events, using historical data.
- Predictive Analytics provides insight into what will happen in the future.
- Prescriptive Analytics helps with decision-making by providing actionable advice.

Anonymization: The process in which individually identifiable data is altered in such a way that it no longer can be related to a given individual. Among many techniques, there are three primary ways that data is anonymized.

1. Suppression is the most basic version of anonymization, and it simply removes some identifying values from data to reduce its identifiability.
2. Generalization takes specific identifying values and makes them broader, such as changing a specific age (18) to an age range (18–24).

3. Noise addition takes identifying values from a given data set and switches them with identifying values from another individual in that data set. Note that all these processes will not guarantee that data is no longer identifiable and have to be performed in such a way that does not harm the usability of the data.

Anthropomorphism: The tendency for people to attribute human-like qualities or characteristics to an AI chatbot.

Application Programming Interface (API): A software contract between the application and client, expressed as a collection of methods or functions... it defines the available functions you can execute, the intermediary interface between the client, and the application.

Artificial General Intelligence: A non-human mechanism that demonstrates a broad range of problem solving, creativity, and adaptability. For example, a program demonstrating artificial general intelligence could translate text, compose symphonies, and excel at games that have not yet been invented.

Artificial Intelligence or AI systems: Artificial intelligence (AI) systems are software (and possibly also hardware) systems designed by humans that, given a complex goal, act in the physical or digital dimension by perceiving their environment through data acquisition, interpreting the collected structured or unstructured data, reasoning on the knowledge, or processing the information, derived from this data and deciding the best action(s) to take to achieve the given goal. AI systems can either use symbolic rules or learn a numeric model, and they can also adapt their behavior by analyzing how the environment is affected by their previous actions.

As a scientific discipline, AI includes several approaches and techniques, such as machine learning (of which deep learning and reinforcement learning are specific examples), machine reasoning (which includes planning, scheduling, knowledge representation, reasoning, search, and optimization), and robotics (which includes control, perception, sensors, and actuators, as well as the integration of all other techniques into cyber-physical systems).[8]

Artificial Intelligence Algorithm: A set of instructions or computations that a machine follows in order to learn how to do a particular task.

Artificial Intelligence Alignment: A field of AI safety research that aims to ensure artificial intelligence systems achieve desired outcomes.

Artificial Intelligence as a Service (AIaaS): It is the third-party offering of artificial intelligence (AI) outsourcing. It enables individuals and companies to experiment with AI for various purposes without a large initial investment and with lower risk.

Artificial Intelligence Assurance: The defensive science of protecting artificial intelligence applications from attack or malfunction.

Artificial Intelligence Bias: See Machine Learning Bias.

Artificial Intelligence Ethics: Is a subfield of applied ethics, focusing on the ethical issues raised by the development, deployment, and use of AI. Its central concern is to identify how AI can advance or raise concerns for the good life of individuals, whether in terms of quality of life, or human autonomy and freedom necessary for a democratic society.

Artificial Intelligence Governance: Is the legal framework for ensuring AI and machine learning technologies are researched and developed with the goal of helping humanity navigate the adoption and use of these systems in ethical and responsible ways. AI governance aims to close the gap that exists between accountability and ethics in technological advancement.

Artificial Intelligence Learning: The ingestion of a corpus, application of semantic mapping, and relevant ontology of structured and/or unstructured data that yields inference and correlation leads to the creation of useful conclusive or predictive capabilities in a given knowledge domain. Strong AI learning also includes the capability of creating unique hypotheses, attributing data relevance, processing data relationships, and updating its lines of inquiry to further the usefulness of its purpose.

Artificial Intelligence Principles: An overarching concept, value, belief, or norm that guides AI development, testing, and deployment across the AI lifecycle. The OECD identifies five complementary values-based principles for the responsible stewardship of trustworthy AI and calls on AI actors to promote and implement them:

- Inclusive growth.
- Sustainable development and well-being.
- Human-centered values and fairness.
- Transparency and explainability, robustness, security, and safety.
- Accountability.

Artificial intelligence Prompt: Is a mode of interaction between a human and a large language model that lets the model generate the intended output. This interaction can be in the form of a question, text, code snippets, or examples.

Artificial Intelligence Prompt Engineer: An expert in creating text-based prompts or cues that can be interpreted and understood by large language models and generative AI tools. In contrast to traditional computer engineers who write code, prompt engineers can use written language to evaluate AI systems for idiosyncrasies.

Artificial Intelligence Safety: A field of research concerned with how AI can be developed and used safely and ethically. Issues discussed by AI safety researchers include alignment, bias, the disruptive effects of automation, hallucination, and the use of generative AI to create deepfakes and misinformation.

Artificial Intelligence Security: Refers to the measures that one can take to protect AI systems from cyber-attacks, data breaches, and other security threats.

Artificial Intelligence System: An engineered system that generates outputs, such as content, forecasts, recommendations, or decisions for a given set of human-defined objectives. The engineered system can use various techniques and approaches related to artificial intelligence to develop a model to represent data, knowledge, processes, etc. which can be used to conduct tasks. AI systems are designed to operate with varying levels of automation.

Artificial Narrow Intelligence (ANI): Artificial Narrow Intelligence, also known as weak or applied intelligence, represents most of the current artificial intelligence systems which usually focus on a specific task.

Narrow AIs are mostly much better than humans at the task they were made for: for example, look at face recognition, chess computers, calculus, and translation.

The definition of artificial narrow intelligence is in contrast to that of strong AI or artificial general intelligence, which aims at providing a system with consciousness or the ability to solve any problems. Virtual assistants and AlphaGo are examples of artificial narrow intelligence systems.

Artificial Neural Networks (ANNs): A mathematical system, modeled on the human brain, that learns skills by finding statistical patterns in data. It consists of layers of artificial neurons.

Nodes or artificial neurons are each responsible for a simple computation. These nodes are networked together with connections of varying strengths, and learning is reflected in changes to those connections.

An important characteristic of neural networks is the relationship between nodes. Often, there is an input layer, an output layer, and one or more in between layers (called "hidden layers"), which can result in a model that has a lot of complexity but may be difficult to interpret.

AI System's Lifecycle: The end-to-end process of designing, developing, deploying, monitoring, and retiring an AI system. It includes stages such as problem identification, data collection, model training, testing, deployment, maintenance, and eventual decommissioning.

Auditability: Auditability refers to the ability of an AI system to undergo the assessment of the system's algorithms, data, and design processes. This does not necessarily imply that information about business models and Intellectual Property related to the AI system must always be openly available. Ensuring traceability and logging mechanisms from the early design phase of the AI system can help enable the system's auditability.[9]

Adversarial Attacks: Refer to deliberate manipulations of input data designed to deceive machine learning models. These attacks exploit vulnerabilities in the model by introducing subtle perturbations – often imperceptible to humans – that cause the model to produce incorrect predictions or classifications. Adversarial attacks are commonly studied in the field of AI security and can target various models, including image recognition systems, natural language processing systems, and autonomous systems, highlighting the need for robust defenses.

Adversarial Evasion: A type of adversarial attack in the field of AI where an attacker modifies input data to bypass a machine learning model's detection or classification without altering the underlying content significantly.

The goal is to "evade" the system by crafting adversarial examples – inputs specifically designed to exploit the model's vulnerabilities – so that the system misclassifies them during inference. Adversarial evasion is commonly studied in contexts like malware detection, spam filters, and image recognition systems, where attackers aim to avoid detection or scrutiny.

Adversarial Inference: A technique used by attackers to extract sensitive information about a machine learning model or its training data. This technique specifically involves querying a model to deduce information about its structure, parameters, or decision boundaries, often to replicate or exploit it.

Adversarial Inversion: A technique used by attackers to extract sensitive information about a machine learning model or its training data. This technique specifically targets the recovery of input data, such as reconstructing sensitive training samples (e.g., images, text, or personal information) from the model's outputs or gradients.

Adversarial Model Extraction: An attack where an adversary aims to replicate or approximate a target machine learning model by querying it and analyzing its outputs. The attacker uses these queries to infer details about the model, such as its architecture, hyperparameters, or learned weights, effectively reconstructing a copy of the model. This can compromise intellectual property, allow unauthorized access to the model's capabilities, or facilitate further adversarial attacks, such as evasion or inversion. Model extraction is a significant concern in scenarios involving proprietary or sensitive AI systems deployed via APIs or services.

Adversarial Poisoning: An attack in the field of AI where an adversary corrupts the training data of a machine learning model to degrade its performance or manipulate its behavior. This type of attack involves injecting malicious data points into the training set or altering existing ones, causing the model to learn incorrect patterns or biases. Adversarial poisoning can result in compromised models that make inaccurate predictions or behave in ways desired by the attacker, and it is a significant concern in environments where data integrity cannot be fully controlled, such as? .

Adversarial Robustness: The ability of a machine learning model to function correctly when confronted with an external agent trying to fool it.

Adversarial Success: Indicates reaching an availability breakdown, integrity violations, privacy compromise, or abuse trigger (for generative AI models only) in response to attempted adversarial attacks on the model.

Authentication: "An act, process, or method of showing something to be real, true, or genuine" and in the case of AI, data, models, and outputs, the act of verifying or confirming the authenticity of content generated by AI models.

Bias: Systematic deviation in AI outputs that reflects or amplifies unfair or unintended preferences or prejudices in the data, model, or deployment environment, potentially leading to discrimination or incorrect predictions.

Bias Mitigation: Taking steps to reduce or eliminate unfair and prejudiced influences in data or algorithms to ensure that AI systems make decisions without favoring or disadvantaging specific groups.[10]

Bias Mitigation Algorithm: A procedure for reducing unwanted bias in training data or models.[11]

Black Box: The nature of some AI techniques whereby the inferential operations are complex, hidden, or otherwise opaque to their developers and end-users in terms of providing an understanding of how classifications, recommendations, or actions are generated and what overall performance will be.

Catastrophic Forgetting: Catastrophic forgetting, or catastrophic interference, is a phenomenon in neural networks where learning new information disrupts or erases previously learned information. This is a significant challenge in continual learning scenarios, where the model needs to retain knowledge from past tasks while learning new ones.

Chain-of-Thought Prompting: Chain-of-thought prompting is a technique used in natural language processing where an AI model is guided to break down complex reasoning tasks into a series of smaller, logical steps. This approach helps enhance the model's performance on tasks that require multi-step reasoning by explicitly outlining the intermediate steps in the thought process.

Chatbots (or "bots"): A program designed to help human users with simple tasks and communicate via voice or text commands to feel like a human-to-human conversation.

ChatGPT: ChatGPT (short for Chat Generative Pre-trained Transformer) is a chatbot launched by OpenAI in November 2022. It is a Large Language Model with both supervised and reinforcement learning techniques. ChatGPT can produce a body of unique text from a user's specific input based on existing content from the internet.

Compliance Audit: An independent review of an organization's adherence to regulatory guidelines.

Confabulation: In the context of artificial intelligence, confabulation refers to the generation of plausible but fabricated or distorted information by an AI system, particularly in situations where it lacks sufficient knowledge or data. This often happens in generative models when they create outputs that sound credible but are not based on real or accurate information.

Data Curator: An individual whose expertise is in finding the right datasets and leveraging data marketplaces to acquire good data. They need to interact with legal, finance, IT, and product teams (data scientists and engineers) for both buying or selling data and model assets.[12]

Data Leakage: Occurs when information from outside the designed training dataset is used in the creation of the model, contaminating the training dataset.[13]

Data Poisoning: The method used to manipulate or distort datasets and AI models by inserting corrupted information or data into the model training process. If performed maliciously, the objectives of data poisoning are primarily to trick models to bias their outputs.

Data Provenance: Refers to a record that accounts for the origin of a piece of data (in a database, document, or repository), together with an explanation of how and why it got to the present place. A provenance record will document the history of each piece of data.[14]

Datasets: A collection of information that a machine learning algorithm uses to "learn" patterns to use for its predictions.

Data Sketching: Refers to a process of creating compact, approximate representations of large datasets using algorithmic techniques to enable efficient analysis, computation, or storage. It is commonly used for tasks like summarizing data streams, estimating patterns, or identifying trends without processing the full dataset.

Datasheet: Is a structured document that provides detailed metadata and documentation about a dataset, including its creation, purpose, composition, limitations, and ethical considerations. It aims to ensure transparency, accountability, and proper usage of the dataset throughout its lifecycle.

Deepfakes: The result of distorting or manipulating AI models or data to generate convincing image, audio, or video content that appears real but is in fact misrepresentative or false.

Disinformation: False information that is spread with the intent to mislead individuals into believing it is true.

ELI5 (Explain Like I'm 5): A Python library that allows you to visualize and debug various machine learning models using a unified API. It has built-in support for several popular machine learning libraries and provides a way to compute feature importance for any black-box estimator by measuring feature importance using feature permutation.[15]

Ethics: Principles and standards governing what is considered morally right or wrong in the development and deployment of AI systems, emphasizing fairness, accountability, and respect for human rights.

Ethical AI: Designing, developing, and deploying AI systems that align with ethical principles, such as fairness, accountability, transparency, and inclusivity, ensures they benefit society and avoid harm.

Explainability: A characteristic of an AI system is that there is provision of accompanying evidence or reasons for system output in a manner that is meaningful or understandable to individual users (as well as to developers and auditors) and reflects the system's process for generating the output (e.g., what alternatives were considered, but not proposed, and why not).

FactSheet: A collection of key facts about how an AI model or service* was developed, tested, deployed, monitored, and modified over time that is rendered in a way that is understandable to the consumer.[16]

Fairness Metric: A quantification of unwanted bias in training data or models.[17]

Federated Learning: A learning model that addresses the problem of data governance and privacy by training algorithms collaboratively without transferring the data to another location.

Few-Shot Prompting: Refers to a technique where a model is trained to perform tasks or generate responses with only a small number of examples or prompts, typically a handful, rather than relying on extensive training data. This approach demonstrates the model's ability to generalize and adapt quickly to new tasks or questions based on very limited input.

General Adversarial Networks (GANs): GANs are a class of machine learning frameworks where two neural networks, the generator and the discriminator, are trained simultaneously through adversarial processes. The generator creates fake data samples to mimic real data, while the discriminator evaluates them. The goal is for the generator to produce data indistinguishable from real data, thereby improving its ability to generate realistic samples over time.

Generative AI: A type of artificial intelligence that can create or produce different types of content, including text, imagery, audio, and synthetic data based on patterns and predictions in the data.

Gini Importance (Random Forest): Also known as Mean Decrease Impurity. It is a method used in random forest models. It calculates feature importance as the total decrease in node impurities from splitting on the feature, averaged over all trees in the model.

The advantage of this method is that it considers interaction effects and can handle different types of features. However, it can be biased toward features with more categories and might overestimate the importance of features being correlated.[18]

Governance: The actions to ensure stakeholder needs, conditions, and options are evaluated to determine balanced, agreed-upon enterprise objectives; setting direction through prioritization and decision-making; and monitoring performance and compliance against agreed-upon directions and objectives. AI governance may include policies on the nature of AI applications developed and deployed versus those limited or withheld.

Hallucination: When an AI model generates false or misleading information but presents it as if it were a fact and/or truthful.

Hugging Face: It is a machine learning (ML) and data science platform and community that helps users build, deploy, and train ML models. It provides the infrastructure to demo, run, and deploy artificial intelligence (AI) in live applications.

Hyperparameter: Hyperparameters are used to control the learning process in machine learning. These parameters are controlled by the coder, unlike the values of the model parameters which are derived via training.[19]

Individual Conditional Expectation (ICE) plots: ICE plots show how much a predicted outcome depends on a specific feature. They are similar to PDPs but show the relationship between a feature and the model's output for individual instances, rather than averaging across the dataset. They can complement PDPs by providing a more detailed view of the model's behavior – for example, by highlighting variability and showing interactions between features at an instance level.[20]

Intelligent Agent: An intelligent agent is an AI system that can independently perceive its environment and act autonomously to reach an objective. The environment may be simulated or physical and does not need repeated prompting to achieve a bigger task.

Internal Audit: Refers to a specific team that is organizationally independent from senior management. Internal audit as a specific function has a precisely defined scope (evaluating the organization's risk management practices), purpose (providing independent and objective assurance to the board), and position within the organization (it is organizationally independent of senior management and reports directly to the board).

This team performs an ongoing assessment of the organization's risk management practices and reports its findings directly to the board of directors, especially if they discover ineffective practices. In doing so, the board of directors gets more objective information about the organization's risk management practices.[21]

Key Performance Indicators (KPIs): These are quantifiable business metrics that corporate executives and other managers use to track and analyze factors deemed crucial to the success of an organization.

Knowledge engineering: It is the field of AI that aims to emulate a human expert's knowledge in a certain field.

Large Language Model (LLM): LLMs are a type of foundation model that can generate text after being trained on massive amounts of unlabeled, textual data.

Local Interpretable Model-Agnostic Explanations (LIME): LIME helps to explain a model's predictions by focusing on a single prediction at a time. It does this by creating a simpler, interpretable model that mimics the behavior of the complex model for that specific prediction.[22]

Machine Learning (ML): ML is a field of computer science dealing with methods to develop (learn) rules from input data to achieve a given goal. A ML algorithm is a certain programmatic implementation of a strategy to find such rules (e.g., with a neural network, a logistic regression, or a decision forest). A ML model is the resulting set of rules (encoded in model parameters), inherently, including the type of the ML algorithm. The ML model can be used to make predictions on data previously unknown to the ML model.[23]

Natural Language Processing: An interdisciplinary subfield of artificial intelligence and linguistics that provides computers with the ability to comprehend text and speech.

Neural Networks: A form of machine learning models known for their ability to learn complex patterns. They consist of several layers of nodes, where each node is a simple, non-linear processing unit.[24]

Overfitting: In statistics and machine learning, overfitting occurs when a model tries to predict a trend in data that is too noisy. Overfitting is the result of an overly complex model with too many parameters.

A model that is overfitted is inaccurate because the trend does not reflect the reality of the data. An overfitted model is a model with a trend line that reflects the errors in the data that it is trained with, instead of accurately predicting unseen data.

Partial Dependence Plots (PDPs): PDPs show how a feature affects the model's predictions, on average, across the dataset. They help visualize the relationship between a feature and the model's output, holding all other features constant.[25]

Prompt Engineering: Refers to the process of carefully designing and crafting the input prompts or queries given to an artificial intelligence system, such as a language model, to elicit desired and specific responses. It involves fine-tuning the wording, structure, and context of prompts to achieve accurate and relevant outputs from the AI system.

Q-learning: Q-learning is a type of reinforcement learning that enables AI models to learn and improve iteratively over time.

Reinforcement Learning (RL) Models: These are a type of machine learning model where an AI agent learns to make decisions by interacting with an environment and receiving feedback in the form of rewards or penalties. RL models aim to maximize cumulative rewards over time by learning the optimal sequence of actions to achieve a specific goal.

Reproducibility: The ability to consistently replicate the results of an AI model or system using the same data, code, and computational environment, ensuring reliability and scientific validation.

Responsible AI: Responsible AI refers to the practice of designing, developing, and deploying AI systems in a manner that is ethical, transparent, and aligned with societal values. It encompasses principles such as fairness, accountability, transparency, and privacy, ensuring that AI technologies are used for the benefit of all individuals and do not cause harm or exacerbate biases.

Rewards: The performance of a behavioral policy is judged by the reward function, which also serves as the directing force for learning (i.e., updating the policy). Rewards may be given by simple, direct measurements (e.g., "time spent watching a video" for a recommendation algorithm), a complex combination of measurements from the state and actions (e.g., lane position and fuel usage in autonomous vehicles), or even another learned model from a previously logged set of behaviors (e.g., an autonomous vehicle attempting to imitate previous trajectories). The system receives a reward signal at each time step, and the sum of these individual rewards constitutes the cumulative reward (though there may be long periods where the system receives zero rewards until completing some task). A common formulation in RL is to weigh near-term rewards more heavily than long-term rewards. This process, known as discounting, involves optimizing a sum of weighted rewards, where the importance of the reward at each time step is reduced by a constant factor known as the discount rate. For example, an RL-trained robot learning to run could be rewarded for every second it does not fall over, rather than for performing a specific sequence of actions, to encourage behavioral exploration.[26]

Risk: When considering the negative impact of a potential event, the risk is a function of (1) the negative impact, or magnitude of the harm, that would arise if the circumstance or event occurs and (2) the likelihood of occurrence. Negative impact or harm can be experienced by individuals, groups, communities, organizations, society, the environment, and the planet.

Robust AI: AI systems that can maintain performance and reliability under diverse and unexpected conditions, such as adversarial attacks, noisy data, or hardware failures.

Scikit-learn: A popular machine learning library in Python. It provides several methods to calculate feature importance. For instance, in the case of decision trees, random forests, and gradient-boosting machines, it calculates feature importance based on the mean decrease in impurity (MDI). For linear and logistic regression, coefficients can be used to determine feature importance.[27]

Shapley Additive Explanations (SHAP): SHAP is a cooperative game theory-style approach to interpretability that considers all possible combinations of features and how they affect the prediction. It assigns a value (called a Shapley value) to each feature based on how much it contributes to the prediction in different scenarios.[28]

Single-Variable Prediction: Also known as univariate feature selection, is a method where you create a simple model for each feature, using that feature alone to make predictions. The performance of these models indicates the importance of each feature.

The advantage of this method is its simplicity and ease of interpretation. However, it fails to capture interactions between features and can be biased if the number of features is large compared with the number of observations.[29]

Singularity: In technology, the singularity describes a hypothetical future where technology growth is out of control and irreversible.

Supplier's Declaration of Conformity (SDoC): Is a formal document provided by a manufacturer, supplier, or distributor stating that a product complies with relevant regulatory, safety, or technical standards. It is typically issued based on self-assessment, supported by appropriate testing or documentation, and does not necessarily require third-party certification. The SDoC ensures that the product meets the specified requirements for sale and use within a specific region or industry.

Target Leakage: This error occurs when a model is trained with information that will not be available in the same way or with the same quality when the model is put into production. This generally has to do with the temporality of the data or groupings. In more subtle cases, this error can be difficult to detect since the variable is present, but the information is updated retroactively.[30]

Traceability: The ability to track and document the development, decision-making processes, and outputs of an AI system throughout its lifecycle to ensure transparency, accountability, and compliance with regulations.

Training Data: A dataset from which a model is learned.

Train, Validation, and Test Data: Initially training data is used to fit an optimal performing ML model. The validation data is then used to provide an evaluation of the model fit while tuning the hyperparameters. The test data is used last to provide a final evaluation of the model's performance.

Transformer: A procedure that modifies a dataset.

Transparent AI: Creating artificial intelligence systems in a way that their operations and decision-making processes are clear and understandable, promoting accountability and trust among users.[31]

Tree-of-Thought Prompting: Tree-of-thought prompting is a method in which the AI model explores multiple potential reasoning paths or decision trees to arrive at a solution. This approach allows the model to consider various possibilities and systematically evaluate them, enhancing its ability to handle complex, multifaceted problems.

Trust: Confidence in an AI system's ability to perform as intended, reliably, and ethically, based on its transparency, fairness, accountability, and performance.

Trust and Transparency (in AI): Refers to the principles and practices that aim to build confidence, understanding, and accountability in AI systems. They address the need for users, stakeholders, and society as a whole to trust and have visibility into how AI technologies operate and make decisions.

Trustworthy AI: AI systems that are designed, developed, and deployed to be reliable, ethical, secure, transparent, and accountable, ensuring they are fair, non-discriminatory, and aligned with societal values.

Turing Test: A test that examines a machine's ability to pass for a human, specifically in language and behavior. A machine can pass this test if, after being graded by a human, its performance is indistinguishable from that of human participants. This test is named after Alan Turing (1912–1954), a notable English mathematician, computer scientist, and logician.

Underfitting: Occurs when the model is unable to perform well with the training data or generalize to new data. This happens when individual characteristics of observations are over-grouped and given little weight. An underfitted model tends to ignore patterns in the predictive structure.

This is reflected in systematic and identifiable errors – for example, systematic under or overprediction for certain groups or values of the input variables.

Uncertainty Quantification (UQ): Gives AI the ability to express that it is unsure, adding critical transparency for the safe deployment and use of AI.[32]

Validation Set: Is a set of data used to train artificial intelligence (AI) with the goal of finding and optimizing the best model to solve a given problem. Validation sets are also known as dev sets.

White Box AI: Also known as transparent AI or explainable AI, refers to artificial intelligence systems or models whose internal workings and decision-making processes are understandable and interpretable by humans. In white box AI, the algorithms and mechanisms used to arrive at a particular output or decision can be traced, visualized, and comprehended, providing insights into the factors that influenced the AI's outcome.

XGBoost (eXtreme Gradient Boosting): A popular implementation of the gradient-boosting decision trees algorithm is used for classification, regression, and ranking problems. XGBoost provides a straightforward way to calculate feature importance based on the number of times a feature appears in a tree across all the trees in the model. It also provides another method known as "gain," which calculates the average gain of a feature when it is used in trees.[33]

NOTES

1 The International Federation of Red Cross and Red Crescent Societies (IFRC), 2023, "What is a disaster?," www.ifrc.org/our-work/disasters-climate-and-crises/what-disaster and United Nations Office for Disaster Risk Reduction (UNDRR), 2023, "Disaster: Sendai Framework Terminology On Disaster Risk Reduction, www.undrr.org/terminology/disaster.

2 (n.a.), January 26, 2024, "Ai Fairness," Clan, https://clanx.ai/glossary/ai-fairness

3 OECD, May 2024, "Defining AI incidents and related terms" OECD Artificial Intelligence Papers, No. 16, OECD Publishing, Paris, www.oecd-ilibrary.org/deliver/d1a8d965-en.pdf?itemId=%2Fcontent%2Fpaper%2Fd1a8d965-enandmimeType=pdf

4 IBIB.

5 High-Level Expert Group on Artificial Intelligence, April 8, 2019, "Ethics Guidelines for Trustworthy AI," European Commission, https://digital-strategy.ec.europa.eu/en/library/ethics-guidelines-trustworthy-ai

6 TrustArc, February 19, 2024, "Responsible AI Certification Assessment Criteria, v1.0," https://trustarc.com/wp-content/uploads/2024/03/Responsbile-AI-Certification-Assessment-Criteria.pdf.

7 Verma, S., and Rubin, J. (2018). Fairness Definitions Explained. Conference Proceedings of the International Workshop on Software Fairness (págs. 1-7). New York, NY, USA: Association for Computing Machinery, https://fairware.cs.umass.edu/papers/Verma.pdf.

8 Verma, S., and Rubin, J. (2018). Fairness Definitions Explained. Conference Proceedings of the International Workshop on Software Fairness (páges 1–7). New York, NY, USA: Association for Computing Machinery, https://fairware.cs.umass.edu/papers/Verma.pdf.
9 Verma, S., and Rubin, J. (2018). Fairness Definitions Explained. Conference Proceedings of the International Workshop on Software Fairness (págs. 1-7). New York, NY, USA: Association for Computing Machinery, https://fairware.cs.umass.edu/papers/Verma.pdf.
10 (n.a.), January 26, 2024, "Ai Fairness," Clan, https://clanx.ai/glossary/ai-fairness
11 (n.a.), (n.d.), "AI Fairness 360 – Resources," IBM Research Trusted AI, https://aif360.res.ibm.com/resources#glossary.
12 Voruganti, K., Steele, P., April 16, 2024, "Tapping into AI Model Marketplaces to Unlock Value: Part 1," Equinix, https://blog.equinix.com/blog/2024/04/16/tapping-into-ai-model-marketplaces-to-unlock-value-part-1/#_edn2.
13 Kaufman, S., Rosset, S., and Perlich, C., December 1, 2011, "Leakage in Data Mining: Formulation, Detection, and Avoidance," ACM Transactions on Knowledge Discovery from Data (TKDD), Volume 6, Issue 4, Article No.: 15, Pages 1–21, www.cs.umb.edu/~ding/history/470_670_fall_2011/papers/cs670_Tran_PreferredPaper_LeakingInDataMining.pdf.
14 Ariga, T., Persons, T., Sanford, S., June 2021, "Artificial Intelligence Accountability Framework, An Accountability Framework for Federal Agencies and Other Entities," U.S. Government Accountability Office (GAO), GAO-21-519SP, www.gao.gov/assets/gao-21-519sp.pdf.
15 Kolena Editorial Team, 2024, "Feature Importance: Methods, Tools, and Best Practices,' Kolena, www.kolena.com/guides/feature-importance-methods-tools-and-best-practices-2.
16 (n.a.), (n.d.), "AI Fairness 360 – Resources," IBM Research Trusted AI, https://aif360.res.ibm.com/resources#glossary
17 (n.a.), (n.d.), "AI Fairness 360 – Resources," IBM Research Trusted AI, https://aif360.res.ibm.com/resources#glossary.
18 Kolena Editorial Team, 2024, "Feature Importance: Methods, Tools, and Best Practices,' Kolena, www.kolena.com/guides/feature-importance-methods-tools-and-best-practices-2.
19 Supreme Audit Institutions of Finland, Germany, the Netherlands, Norway, and the UK, April 7, 2023, "Auditing machine learning algorithms," www.auditingalgorithms.net.
20 McGrath, A., Jonker, A., October 8, 2024, "What is AI interpretability?" IBM, www.ibm.com/think/topics/interpretability.
21 Schuett, J. (2024). Frontier AI developers need an internal audit function. Risk Analysis, 1–21. https://doi.org/10.1111/risa.17665.
22 McGrath, A., Jonker, A., October 8, 2024, "What is AI interpretability?" IBM, www.ibm.com/think/topics/interpretability.
23 Supreme Audit Institutions of Finland, Germany, the Netherlands, Norway, and the UK, April 7, 2023, "Auditing machine learning algorithms," www.auditingalgorithms.net.
24 Kolena Editorial Team, 2024, "Feature Importance: Methods, Tools, and Best Practices,' Kolena, www.kolena.com/guides/feature-importance-methods-tools-and-best-practices-2.

25 McGrath, A., Jonker, A., October 8, 2024, "What is AI interpretability?" IBM, www.ibm.com/think/topics/interpretability.
26 Gilbert, T., Lambert, N., Dean, S., and Zick, T., February 2022, "Choices, Risks, and Reward Reports: Charting Public Policy for Reinforcement Learning Systems," UC Berkeley, Center for Long-Term Cybersecurity, Artificial Intelligence Security Initiative (AISI), https://cltc.berkeley.edu/wp-content/uploads/2022/02/Choices_Risks_Reward_Reports.pdf.
27 Kolena Editorial Team, 2024, "Feature Importance: Methods, Tools, and Best Practices,' Kolena, www.kolena.com/guides/feature-importance-methods-tools-and-best-practices-2.
28 McGrath, A., Jonker, A., October 8, 2024, "What is AI interpretability?" IBM, www.ibm.com/think/topics/interpretability.
29 Kolena Editorial Team, 2024, "Feature Importance: Methods, Tools, and Best Practices,' Kolena, www.kolena.com/guides/feature-importance-methods-tools-and-best-practices-2.
30 Kaufman, S., Rosset, S., and Perlich, C., December 1, 2011, "Leakage in Data Mining: Formulation, Detection, and Avoidance," ACM Transactions on Knowledge Discovery from Data (TKDD), Volume 6, Issue 4, Article No.: 15, Pages 1 – 21, www.cs.umb.edu/~ding/history/470_670_fall_2011/papers/cs670_Tran_PreferredPaper_LeakingInDataMining.pdf.
31 (n.a.), January 26, 2024, "Ai Fairness," Clan, https://clanx.ai/glossary/ai-fairness.
32 n.a.), (n.d.), "Uncertainty Quantification 360," IBM Research Trusted AI, https://uq360.res.ibm.com.
33 Kolena Editorial Team, 2024, "Feature Importance: Methods, Tools, and Best Practices,' Kolena, www.kolena.com/guides/feature-importance-methods-tools-and-best-practices-2.

Index

Note: **Bold** page numbers refer to tables; *italic* page numbers refer to figures and page numbers followed by "n" denote endnotes.

For Product Safety Concerns and Information please contact our EU
representative GPSR@taylorandfrancis.com
Taylor & Francis Verlag GmbH, Kaufingerstraße 24, 80331 München, Germany

www.ingramcontent.com/pod-product-compliance
Ingram Content Group UK Ltd.
Pitfield, Milton Keynes, MK11 3LW, UK
UKHW021832260926
475702UK00049B/352

* 9 7 8 1 0 4 1 0 7 8 7 9 1 *